D0174987

PLANT
GEOGRAPHY

*with special reference
to North America*

PHYSIOLOGICAL ECOLOGY

A Series of Monographs, Texts, and Treatises

EDITED BY

T. T. KOZLOWSKI

University of Wisconsin
Madison, Wisconsin

T. T. KOZLOWSKI. Growth and Development of Trees, Volumes I and II – 1971

DANIEL HILLEL. Soil and Water: Physical Principles and Processes, 1971

J. LEVITT. Responses of Plants to Environmental Stresses, 1972

V. B. YOUNGNER AND C. M. MCKELL (Eds.). The Biology and Utilization of Grasses, 1972

T. T. KOZLOWSKI (Ed.). Seed Biology, Volumes I, II, and III – 1972

YOAV WAISEL. Biology of Halophytes, 1972

G. C. MARKS AND T. T. KOZLOWSKI (Eds.). Ectomycorrhizae: Their Ecology and Physiology, 1973

T. T. KOZLOWSKI (Ed.). Shedding of Plant Parts, 1973

ELROY L. RICE. Allelopathy, 1974

T. T. KOZLOWSKI AND C. E. AHLGREN (Eds.). Fire and Ecosystems, 1974

J. BRIAN MUDD AND T. T. KOZLOWSKI (Eds.). Responses of Plants to Air Pollution, 1975

REXFORD DAUBENMIRE, Plant Geography, 1978

PLANT GEOGRAPHY

*with special reference
to North America*

Rexford Daubenmire

ACADEMIC PRESS
New York   San Francisco   London   1978
A Subsidiary of Harcourt Brace Jovanovich, Publishers

ACADEMIC PRESS, INC.
111 Fifth Avenue, New York, New York 10003

United Kingdom Edition published by
ACADEMIC PRESS, INC. (LONDON) LTD.
24/28 Oval Road, London NW1

Library of Congress Cataloging in Publication Data

Daubenmire, Rexford F   Date
    Plant geography.

    (Physiological ecology series)
    Bibliography: p.
    Includes index.
    1. Phytogeography.   2. Botany—North America.
I. Title.
QK101.D26   581.9   77-75570
ISBN 0-12-204150-X

PRINTED IN THE UNITED STATES OF AMERICA

79 80 81 82   9 8 7 6 5 4 3 2

CONTENTS

All royalties from the sale of this book have been donated by the author to The Nature Conservancy.

PLANT GEOGRAPHY

*with special reference
to North America*

INTRODUCTION

Plant geography (or phytogeography, or geobotany) is that branch of science concerned with describing and interpreting the uneven distribution of the earth's plant life. Clearly its origin had to await the birth of at least a primitive type of taxonomy, for only after plants came to have somewhat standardized names could information accumulate regarding the details of distribution, and thus expose patterns that challenge the mind for explanation. Even in the time of Theophrastus, however, there was enough known to elicit commentary on the restrictions of species to different geographic areas. During the Middle Ages nomenclatorial confusion was still a handicap to progress, as evident in comparing the herbals of that period. But when Linnaeus made his epic contribution to standardized nomenclature in the eighteenth century, taxonomic concepts were greatly improved and distribution patterns could be better defined. The stage was then set for serious inquiry into this fascinating field.

Interest at first centered simply upon description from a static point of view, but even as early as 1792, it became apparent to K. L. Willdenow, at least, that floras are not static, that they have attained their present composition and ranges by means of past evolution, dissemination efficiency, migration, and extinction. However, in those times, for fear of torture sponsored by entrenched religions, such a hypothesis had to be expressed cautiously by recognizing "centers of creation." Willdenow, for example, suggested that each major mountain system could have served as such a center, but Christian dogma would not admit the possibility of extinctions.

A century later, as Charles Darwin's dynamic concept of species permeated the fabric of the natural sciences, open assault was begun on the problems of origin, migration, and evolution in plants. If closely related species have evolved from a common ancestor, then a genus has to be older than its species, and in the same way the family still older. Also a species

1

must have had a place of origin from which it spread until it encountered barriers, and if the features of the earth's surface are to be given dynamic interpretation, then old barriers could disappear and new discontinuities arise. Paleobotany, historical geology, and taxonomy began to be drawn closer together by these considerations, and in the present century cytogenetics too has made important contributions toward the solution of problems concerning the origin and spread of taxa.

A different approach was added to the field of plant geography during the nineteenth century when those botanists who became interested in plant communities began to study the distribution of these composite entities in relation to climate. Alexander von Humboldt stands out as the first to recognize physiognomic groups of plants, irrespective of their taxonomic relations, and show that these are geographic units relatable to climate. This shifting of emphasis away from taxonomy toward ecology attracted increasing numbers of workers so that by 1849 J. Thurman called attention to the fact that there had developed two major approaches to plant geography and defined their scope.

One of these is *floristic plant geography*, which is primarily a study of evolutionary divergence, migration, and decline of taxa, as influenced by past events of the earth's history. The basic data here consist of maps of plant ranges and geologic features, fossil floras, and chromosome counts. Differences or similarities in the composition of floras of different regions are determined and given historical interpretation. The relative abundance of species is of little concern, except that special attention is accorded rare species. Neither the physiognomy of plant life, nor the dynamic interrelations among species, is considered. In this endeavor the disciplines of geology, paleontology, morphology, and cytogenetics are employed more often in making explanations than is autecology.

The alternative approach to plant geography, which came to be known as *ecologic plant geography*, takes plant communities as units having ranges to be interpreted. Here the outlook is dominated by sociologic and physiologic, rather than phylogenetic and historic considerations. Explanations of the boundaries of vegetation units are sought on the basis of structural and functional adaptations to various environmental complexes, without laying particular stress on the geographic position of the region. The problem of how and when a group of ecologically related plants came to be where it is at present is usually considered of minor importance as compared with the problem of how this community is able to maintain its space relations in the present pattern of environments. Convergent evolution is the chief aspect of paleobotany of concern to the ecologic plant geographer, for he is acutely aware that under similar environments taxonomically diverse floras may produce communities having ecologic and physiognomic similarity through adaptive evolution. Also, plant succession leading to the closest possible fit of community composition to each combination of environmental factors is a matter of prime importance to ecologic plant geography, whereas it is of no particular significance in floristic plant geography. Abundant and regularly present species that determine the distinctive character of each vegetation unit get more attention than the rare ones.

Detailed quantitative studies of existing vegetation and measurements of present environmental conditions can be subjected to endless tests, therefore all problems of ecologic plant geography are theoretically capable of solution. The great handicaps here are the complexity of vegetation, and the complexity of environment, including problems created by man's activities in the recent past.

Floristic plant geography is not burdened with these complexities, but it usually necessitates tenuous hypotheses concerning the remote past, many of which cannot be proven. For each problem in this field there are several to many plausible explanations, and such can only be rated as to their probable correctness, as best man can judge. Taxonomic concepts are vital, and these are subject to frequent revision. Accurate distribution maps are essential, but most maps are only crude approximations. There is no way of determining accurately the age of a taxon, the absolute limits of its effective dissemination, or whether it is now expanding or contracting its range. These problems lie at the core of floristic plant geography.

Diverse as these two viewpoints seem and as diversely as they have been treated, especially in North America, they are really mutually supporting aspects of the broader field of plant geography. No segment of the earth's green mantle can be well understood without recourse to both types of inquiry. Although books on plant geography vary greatly as to their content, most of them have been largely, if not exclusively, devoted to floristic plant geography, with little or no attention given to a vegetation hierarchy related to climate, soil, and disturbance, i.e., to ecologic plant geography. Both aspects of the subject are considered in the treatment that follows, although not with equal emphasis. Only a concise summary is given of some of the main concepts of floristic plant geography, with most of the space devoted to ecologic plant geography, and in both parts interest is centered on the northern half of the Western Hemisphere. In addition to this special combination and direction of emphasis, more than usual consideration has been given to the probable geologic history of each vegetation unit, as well as this can be surmised from the present stock of paleontologic and geologic information.

Only international names have been used for plants as a means of saving space and avoiding confusion. Most persons would like to imagine that each species has but one colloquial name, but many have none (e.g., in *Carex, Panicum, Quercus, Rubus, Salix.*) and most have several. Even in western North America, where it is endemic, *Purshia tridentata* is called "antelope brush" in places, and "bitterbrush" elsewhere. The "beargrass" of the hot Texas plains is in a different genus than the "beargrass" of cold ridgetops in the Cascade Mountains, or the "beargrass" of Florida sands, and none is a grass. "Ironwood" refers to trees in different families in Arizona, Indiana, and Florida, etc. Since Mexico is considered to the extent which the status of vegetation research there permits, would English-speaking readers take kindly to colloquial Spanish names and vice versa, despite the equal validity of the English and Spanish names?

Infraspecific taxa are commonly indicated as trinomials, and in places I have used trinomials (even names considered out of date) instead of widely accepted binomials, simply to point up significant taxonomic relationships otherwise unrecognizable. Unless species names are alphabetized in listing, there is a connotation of their characteristic order of relative abundance or ubiquity. A genus name used alone usually implies more than one species.

Nearly every term that could possibly be used in describing and interpreting vegetation has been used in different senses, usually with little or no attempt to clarify the special meaning attached to it by each author. It seems futile to try to cast aside terms owing to multiple definitions, so I have selected those that appear to be most meaningful in the English language, and appended a glossary to indicate the particular meaning attached here.

The rod that appears for scale in many of my photographs is 1 m in length, marked off in tenths.

Part I

FLORISTIC PLANT GEOGRAPHY

Floristic plant geography becomes a valid field of inquiry only if one accepts the thesis that the ranges of plants have changed through time. To appraise the validity of this widely held thesis, we shall consider the supporting evidence, but first some basic terms and concepts need clarification.

RELATION BETWEEN DISSEMINATION AND MIGRATION

Within the margin of a species' range,* the scattering of reproductive structures away from each parent plant serves three functions. (1) It minimizes intraspecific competition. (2) It allows new genes to spread through all habitats where the resulting characters are compatible. (3) It tends to extend the range of a species over all the area in which completion of its life cycle is possible.

At the margin of a range, dissemination in all directions from parent plants inevitably puts some disseminules into previously unoccupied territory, as for example, forest tree seeds falling in the margin of contiguous steppe. If the new area permits germination and the subsequent completion of the life cycle, then the process may be repeated with migration extending the range still farther.

Ignoring for the moment such spread as is accomplished by rhizome or runner extension, dissemination is necessary for migration, but by no means does it usually result in migration. Millions of forest tree seeds perish in the margin of contiguous steppe each year.

*See glossary for special terms, or common words used in a special sense.

A species' range may remain under essentially continuous environmental control during migration, as when climate changes slowly enough for migration to keep in step. But range limits may be unrelated to environmental control if the plant represents a new mutation, or if it is an alien plant introduced well within the boundaries of suitable environment and it has not yet had time to exploit all the habitat to which it is adapted, or if the environment has changed rapidly.

Migration is a matter of *range expansion* when more territory is added to the previous range. It is a matter of *range displacement* if part or all of the original range becomes uninhabitable.

Migration may be expected to continue until some environmental factor becomes limiting. This factor is called a barrier. Should there be another tract of favorable territory beyond the barrier, the latter must form a continuous belt wider than the distance which any disseminule of the plant can cross if it is to impede migration indefinitely. Thus only a wide river can be a barrier to a plant with heavy disseminules which do not float. The narrow Isthmus of Panama has kept the mangrove tree *Pelliciera rhizophorae* restricted to the Pacific coast.

Since environmental gradients are numerous and often independent, the limiting factors along different segments of the periphery of a stabilized range are usually different. If inadequate heat limits the poleward distribution of an organism, some other factor must certainly set the equatorward limits, and between these borders oceans, or differences in the amount of seasonal distribution of precipitation, may set the east–west margins of its range.

EVIDENCES OF THE DYNAMIC CHARACTER OF RANGES

Since the remodeling of the earth's surface is a never-ending process, environments are continually changing. Old barriers to migration slowly change positions or disappear, and new ones come into being. Ranges tend to keep in adjustment with barriers, but the migration of plants, especially of vascular perennials, is very slow. Each step in migration requires the maturation of a new generation within the dissemination range of individuals which served as the preceding step. Then on the other hand, retractions are slow since once established, most perennials can live out their life span even if conditions that temporarily allowed their establishment cease to exist. Thus, there is often a considerable lag between environmental change and range adjustment. Range extensions and contractions are probably taking place in most parts of the earth at all times, but the rates of migration are so slow and rendered so uncertain by short-term superimposed climatic fluctuations, that man seldom has opportunity to observe significant migrations directly, and much of our knowledge of past migrations must be based on indirect evidence. The dynamic character of plant ranges is indicated by the follow-

ing types of evidence, with other types of evidence of less conclusive character to be brought out later.

Direct Observation

Accidentally or intentionally, man has often transported disseminules
across ocean barriers and then has been able to observe, frequently with
regret, the rapidity with which a plant may extend its range. *Salsola kali*, for
example, was accidentally introduced into central North America from
Eurasia in 1886, and within half a century it had radiated from its point of
introduction to become one of the most common weeds on the continent.
Although direct observation of plant migration has almost entirely concerned
weeds or disease organisms (such as *Cronartium ribicola*, a rust parasite of
Haploxylon pines), through geologic time new introductions with subsequent spread have occurred naturally from time to time.

Fossil Stratigraphy

Pollen preserved in successive layers of a peat deposit show the kinds of
plants that have grown in that vicinity from the time peat began to accumulate up to the present day. In the Ohio Valley many of these profiles show
that *Picea glauca*, *Abies balsamea*, and *Larix laricina* were abundant in that
area soon after Pleistocene (Table 1) ice melted away and peat started accumulating in ponds. Pollens of most of the deciduous angiosperm trees now
abundant in the same area are represented sparingly, if at all, in the bottom
layers but coming up through the strata fossils of coniferous trees dwindle
while those of angiosperms increase, showing clearly that the conifers emi-
grated as the angiosperms immigrated.

Fossils in older sedimentary rocks reveal similar mass migrations of floras
throughout the earth's history, although in less detail than in the younger
strata.

Distribution Patterns

Centers of Origin

For the most part each taxon has originated at one place, which may be
designated as its center of origin. The center may represent a single point
where a rare mutation or hybridization started a new line of descent, or it
may be a sizable area in which evolution changed an entire population
through a period of time. Among mammalian orders, the location of the
earliest fossils in the geologic record has provided substantial evidence of the
location of centers of origin of certain groups, but vascular plants are much
older and their remains are less easily interpreted than are teeth and bones.

TABLE 1

Major Divisions[a] **of Geologic Time, and Some Salient Features of the Earth's History Starting with the Period in Which Angiosperms Achieved Dominance**

Eras	Periods	Epochs	Millions[b] of years since start	Climate	Nature of North American land area
Cenozoic	Quaternary	Holocene	0.012		Elevated and enlarged
		Pleistocene	2.5		Elevated and enlarged
	Tertiary	Pliocene	13.	Progressive cooling and local drying	Cascades and Sierras form, general uplife continues,
		Miocene	26.	Progressive cooling and local drying	Cascades and Sierras form, general uplife continues,
		Oligocene	40.	Progressive cooling and local drying	Rocky Mountains rise
		Eocene	60.	Warmest	Low and restricted
		Paleocene	75.	Cool	Elevated and enlarged
Mesozoic	Cretaceous		144.	Warm	Low and restricted
	Jurassic		180.	(Last period of fern–cycad–conifer dominance)	
	Triassic		225.		
Paleozoic	Permian		270.	(Cycads join ferns and conifers)	
	Carboniferous		350.	(Conifers start to share dominance with ferns)	
	Devonian		400.		
	Silurian		440.	(Ferns, the first land plants, appear)	
	Ordovician		500.		
	Cambrian		600.		
	Pre-Cambrian		4600.		

[a] Other names sometimes used in designating spans of geologic time are:
Paleogene = Paleocene + Eocene + Oligocene.
Neogene = Miocene + Pliocene.
Holocene = Recent + post-Pleistocene.

[b] Estimates, subject to revision.

Reliable fossil evidence indicating the centers of origin of angiosperm families is as yet unavailable. For genera and species many possible criteria for locating centers of origin have been suggested, but most are too uncertain to be useful. A few of these are as follows.

Centers of taxonomic diversity.   If the ranges of species in a genus overlap, and the group is not very widespread, the region of overlap might be suspected as approximating their centers of origin (Fig. 1). Presumably these taxa could have been derived from a common ancestral stock through mutations that have (1) established reproductive incompatibility at once, (2) altered the season of reproduction, or (3) required invasions of new habitats in the environmental mosaic.

On the other hand, if applied to a widespread genus, there is a possibility that after the genus reached a remote area it underwent a burst of evolutionary differentiation. For example, both *Arceuthobium*[182] and *Pinus* are believed to have originated in Asia, but most species by far are in North America.

Another difficulty in applying this concept to widely distributed groups is that the taxonomic work may not be uniform, so that an apparent concentra-

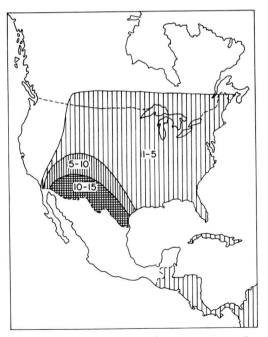

Fig. 1.   Approximate concentration of species in the American genus *Bouteloua*. Only a few species continue southward into South America, so there is an indication of a center of maximum species diversity, and probable center of origin, in the arid belt astride the Mexico–United States border.

tion of species in one area may be largely attributable to narrower concepts of specific or generic limits rather than to greater variability. For example, J. K. Small recognized 44 species of *Cactaceae* in Florida, whereas T. H. Kearney and R. H. Peebles recognized only 72 species in Arizona! Ideally the taxonomy of the group should have been worked out by a single person.

This center of diversity concept seems legitimate when applied to most cultivars, with the most likely source of a cultivar being that geographic area containing the greatest number of closely related wild species; and if this is the true center of origin, longer cultivation in that region should have resulted in the greatest diversity of varieties in the cultivar, adapting it to different soils, agronomic systems, etc.[405]

Trends in evolutionary status.   If the most primitive members of a taxonomic group can be discerned, these may be considered as nearest the geographic center of origin of that group.

Wherever a species or a genus embraces different chromosome numbers which occur in multiples of a common base number, there is little question but that the base number is the most primitive, with successively higher multiples representing successively higher levels of evolution.[96] Thus *Crepis*, containing only as few as four chromosomes in Asia, may well have originated there rather than in the New World where the chromosome numbers of the species are always higher. We cannot be sure of this, however, since the four-chromosome ancestor may have been American which suffered catastrophe in the Western Hemisphere after spreading to Asia.

A more convincing situation has been demonstrated in *Tradescantia*, for several species in this genus have diploids restricted to Texas, with tetraploid populations of the same species more widely ranging. Thus, Texas is strongly indicated as the center of origin of that genus.[10]

Comparative morphology of the species in a genus may reveal a continuous geographic gradient with primitive structures at one extremity and advanced conditions at the other. If the evolutionary interpretation is correct (and this is a crucial point!), the group with primitive characters should be nearest the true center of origin. This technique has indicated Texas as the center of origin of *Lesquerella*.[316]

Reproductive success.   Another possible criterion of center of origin is suggested by the life history of the fungus *Phytophthora infestans*. Although it causes a disease of white potato over wide area, it reproduces sexually only in Mexico.[149] Presumably this pathogen is native to that country, and imperfectly adapted to other areas into which it has followed its host.

Disjunction and Vicariism

When a line is drawn about the outermost localities for a taxon and its range is thus defined on a map, small scale discontinuities that may be

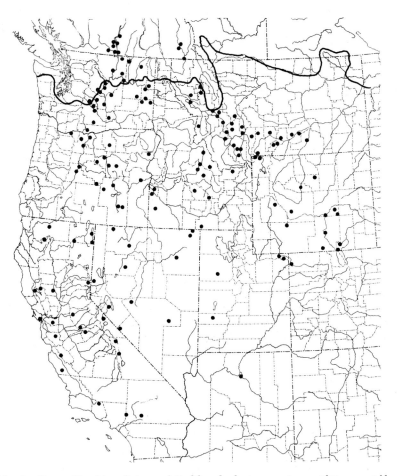

Fig. 2. Range of *Lewisia rediviva* as plotted from herbarium specimens. This perennial herb has obviously migrated north of the glacial border (heavy line) in postglacial time, and judging from its rarity in the south, range displacement is illustrated.

significant for interpreting its past history and present ecology may be obscured. Critical research on distribution clearly requires a map on which all verified localities are indicated by separate dots; the more detailed the base map the better (Fig. 2). (Even in the interpretation of dot maps, the fact must be taken into account that the density of the dots, hence the apparent degree of discontinuity, may reflect primarily ease of access to areas and intensity of collecting!)

Careful study shows that no plants have "continuous" distribution in the absolute sense, for suitable habitats do not occur in unbroken series. However, so long as dissemination or pollination allows gene exchange across unfavorable habitats, directly or indirectly, a range may be said to be continuous despite numerous gaps in which the species cannot be found.

Disjunction may be defined as the occurrence of potentially interbreeding populations separated by a distance exceeding the extreme limits of gene flow. A taxon with disjunct ranges is said to be polytopic (meaning "several places"), or bicentric if only two places are involved.

Some examples of disjunction are as follows. (1) The desert shrub *Larrea divaricata* occurs in one area centered on the Mexico–United States border, and another desert area in northern Argentina. (2) The forest herb *Dicentra cucullaria* occurs in east-central North America, and west of the Rocky Mountains in about the same latitude. (3) The tree *Robinia pseudo-acacia* is native to only the Ozark Mountains, and to the Cumberland Plateau on the opposite side of the broad Mississippi Valley. (4) The fern *Adiantum capillus-veneris* has a population about Fairmont Hot Springs, British Columbia, approximately 900 km north of the remainder of its range.

In 1880 A. R. Wallace published a hypothesis representing a refinement of Willdenow's, namely, that a species arises in one locality, expands its range to a maximum, followed by a decline during which the range becomes disrupted, with fragments dying out one by one until the taxon becomes extinct. Thus, disjunction can be interpreted as a state subsequent to continuous distribution, presumably resulting from vicissitudes that brought about extinction in part of that range. The vast majority of biogeographers consider this model sound, but recognize that it is subject to infinite variations and truncations.

If all individuals of a species form one population, i.e., they can interbreed with each other at least by way of intermediaries, evolution progresses slowly with the population tending to change as a whole. In such a population natural selection favors mutations when they have survival value, while eliminating unfavorable characters promptly. Other innovations are neutral, and these may either be "swamped out" in the large gene pool or may persist to add variety to that gene pool.

On the other hand, if a range should become disjunct, any favorable or neutral innovations that arise subsequently are restricted to the isolated population in which they originated and where they have less overwhelming odds of being swamped out. The smaller the segregate the more rapidly random mutations can become fixed, consequently the more rapidly the population may diverge from the conditions that prevailed before disjunction. The longer the segregates remain out of contact the more likely divergence will proceed to the point where they are no longer compatible with each other. Thus, in the sense of geologic time, disjunction of species tends to be a condition of short duration, for if the disjunct populations persist, divergent evolution tends to convert the segregates into distinct species. With the passing of more time divergence may progress to the genus or family level. When populations attain any degree of taxonomic differentiation, the homologs are called vicariads. Some examples of vicariism are as follows. (1) The genus *Thuja* is represented by different species that are restricted to east-central North America, to the west-central part of the same continent, and to eastern Asia. (2) *Nothofagus* of the Southern Hemisphere and *Fagus* of the Northern Hemisphere are considered as having been derived from a tropical ancestor that is now extinct. (3) Different species of trembling aspen, *Populus tremula* and *P. tremuloides*, occur in Eurasia and North America, respectively.

Vicariism can also involve the same geographic area, i.e., the vicariads are sympatric as opposed to allopatric, if the paired counterparts are limited to different kinds of substrates as with *Fouquieria shrevei* and *F. splendens* on gypsum and nongypsum soils, respectively, or if they occur at different altitudes as with *Artemisia tridentata tridentata* and *A. tridentata vaseyana*. A vast amount of vicariism occurs at the ecotype level, with genetically distinct populations occurring in the same ecosystem mosaic, but separated as a result of different habitat requirements.

One might suspect that the degree of difference between two vicariads would be rather directly proportional to the antiquity of their separation, but wide differences among plants' evolutionary rates keep this from being a useful criterion of age. Annuals are capable of rapid evolution by virtue of their short life spans, so vicariism here could result after short separation. Perennial species can remain stable for long periods of geologic time. Even though they have probably been separated since Miocene time, or at least since a warm interglacial period, no evident differences have developed between disjunct Asiatic and North American populations of *Anaphalis margaritacea, Osmunda claytoni, Phryma leptostachya, Polygonum virginianum, Symplocarpus foetidus,* and *Tovaria virginica*.[175]

Vicariism is but a variation of Wallace's model for interpreting disjunction, in which the disjuncts evolve to varying levels of distinctiveness before becoming extinct. In a number of cases this interpretation has been confirmed by finding fossils in the area separating vicariads or disjuncts. Thus, *Thuja* fossils in Alaska show that this genus was once continuous from North America to Asia. Also, when a group of taxa show the same pattern of discontinuity there is a strong suggestion of former continuity, even where fossils are unknown. As an illustration, many vicariads or even species disjuncts have a tripartite pattern of discontinuity involving east-central North America, west-central North America, and east-central Asia: *Asarum, Coptis, Oplopanax, Menziesia, Thermopsis, Thuja, Trautvetteria, Trillium, Tsuga,* etc. However, where strong evidence of previous continuity is lacking, there are alternative explanations of discontinuity.

Alternative Interpretations of Disjunction and Vicariism

Long-distance dispersal. Disjunction has often been hypothesized to have resulted from a rare accident in which a single disseminule crossed an extensive tract of inhospitable environment and established a population in a new locality. Such an interpretation would be unlikely where disseminules are inefficient, as with the small population of *Carya cordiformis* located in northern Minnesota over 300 km from the margin of continuous populations to the south.[218] Since the seeds of this tree lose their viability upon drying, it is unlikely that aborigines could have carried them to this location in the north.

The crossing of wide area by more mobile disseminules would be most likely to succeed if there were thinly populated habitats in the distant location that would be most receptive to an immigrant. Also long-distance dissemination would more likely be successful if the species were self-compatible. If not self-compatible, the plant would have to be long-lived, or be capable of reproducing itself vegetatively in the new locality until a second individual arrived. Even then in dioecious plants there would be only a 50:50 chance that the second individual would provide the proper sex.[55]

The east coast bulge of Brazil is 2900 km from the west coast of Africa, but a number of plant ranges indicate that transoceanic crossings have been made here. Nearly 2000 species of *Bromeliaceae* are native to the Americas, but only one (*Pitcairnia feliciana*) occurs in the Old World as well as the New, and that species is restricted to the west coast of Africa opposite the bulge of Brazil. Owing to the abundance of bromeliads in South America in contrast to the single disjunction, the situation cannot be interpreted as a consequence of former land continuity. Furthermore, approximately 500 other plants have closely similar patterns of disjunction between South America and Africa, some of the pairs showing the reverse pattern with the more poorly represented member in South America.[391]

The volcanic islands now scattered over the Pacific Ocean have never been connected with any continent, so their native biotas bear testimony that in time many terrestrial species can cross tracts of sea water, if these are not extremely wide. This may be accomplished by "island hopping," by rafting, by violent storms, or by migratory birds. Hawaii is farther from North America (3900 km) than Africa is from Brazil, yet at least 52 taxa have crossed the Pacific Ocean from North America to Hawaii.[143]

Long ago the simplest way of interpreting transoceanic disjunctions was to hypothesize land bridges that arose from the ocean floor, then sank after certain taxa had made the crossing. But geology provided no evidence in support of such hypotheses, and they fell into disrepute. In recent years, however, there have been discovered evidences of islands no longer in existence that could have played the role of stepping stones or "insular isthmuses" which allowed species to cross oceans, either simultaneously or in series. Submarine ridges have been found rising from the ocean floors midway between continents, and volcanic islands on these ridges and on either side of them could have provided land areas spaced closely enough to have served as migratory pathways. Then too, when sea levels were lowered by about 100 m during Pleistocene glaciations, saltwater gaps were narrowed.

Mapping of the Pacific Ocean floor has turned up a great many submerged seamounts, some of which (the "guyots") are flat-topped former volcanic islands as shown by their form and the shallow-water fossils dredged from them. They were eroded down to sea level, then later submerged by subsidence of the ocean floor.[171] Over 160 guyots have been discovered between the Mariana Islands and Hawaii, with their summits mostly 900–1800 m below present sea level. Eleven have been found in the Gulf of Alaska, and two off the coast of southern California. Others occur in the Atlantic Ocean south of the Azores, and in the Caribbean Sea. We must concede that given ample time, limited spans of ocean could have been crossed, even if the journey had to await the successive appearance of different islands. Thus seamounts plus the still existing Pacific islands provide a plausible explanation of the mode of trans-Pacific migration to Hawaii that has long been puzzling. Strengthening this hypothesis, different ecologic types are unequally represented in these islands, with those ill-adapted to even short transport across sea water being poorly represented.

Long ago vascular plants were able to cross freely from North America to Greenland to Iceland to the Faeroes to the Shetlands and the British Isles, or in the reverse direction. Even at present the discontinuities do not exceed 435 km, yet land could not have been continuous across the North Atlantic, for all exchanges of land mammals between North America and Eurasia have been by way of Beringia. Explorations of the Atlantic Ocean floor have recently revealed remnants of a submerged basaltic ridge[187] along this migratory route that vascular plants may have used. Closely spaced projections above sea level could have allowed plants, but not land animals, to have used this migratory tract.

Not all transoceanic disjunctions can be interpreted as migrations involving ocean crossing, even by means of stepping stones. The close floristic affinities between Chile and New Zealand, which share *Nothofagus* and *Araucaria*, seem to have been achieved at some time in the past when these land masses were much closer together, with at least parts of Antarctica extending into temperate latitudes. The fact that neither *Nothofagus* nor *Araucaria* are adapted for dissemination across even narrow bodies of salt water argues strongly for an essentially continuous land connection between South America and New Zealand at some time in the past. Probably these areas drifted apart after the plants had expanded over a single parental continent. *Gyrocarpus americanus*, which is strictly confined to the tropics, reappears in the Americas, Africa, and southern Asia, yet it is inconceivable that such a heavy-fruited land plant could have made crossings of the Atlantic, Pacific, or Indian Oceans even by way of stepping stones. There is nothing to support an alternative hypothesis of parallel evolution from a common ancestor at higher latitudes.

Direct-land connections between two areas would be expected to result in both containing "balanced" floras. That is, there would be approximately the same taxonomic diversity on equivalent areas, both high and low groups would be represented among the taxa, and both widely and narrowly distributed taxa would be present. Unbalanced floras are outstandingly characteristic of remote islands that had to become vegetated by long-distance dissemination.

Transtropical disjunctions are much more controversial than the transoceanic type discussed above. There are about 200 species or species pairs of vascular plants common to west-central North America and southwestern South America, and a hypothesis that the wide discontinuity is a result of long-distance dissemination has been defended by the following facts:[330] (1) The plants are almost without exception self-compatible, a character which favors population establishment by the progeny of a single disseminule. (2) The disjuncts are an unbalanced assemblage of extra tropical families, with many families unrepresented, and others well represented. A land connection should have spread opportunities more evenly among families. (3) No terrestrial vertebrates and only a few insects show the same pattern of discontinuity, which would be unlikely if there had been a continuous strip of land suitable for these plants to follow. (4) A number of the disjuncts require the Mediterranean-type climate, and such has never existed in tropical latitudes. (5) According to fossil evidence, a broad strip of tropical vegetation has intervened between temperate zone floras of the North and South Hemispheres since at least Cretaceous time. (6) Although 29 genera and 10 species extend from one temperate zone to the other along the Andes, even these had to bridge large habitat discontinuities since the Andes uplifted mainly in Pliocene time, and Central America has been a continuous land bridge only since then. (7) Migratory birds, possible vectors, frequently fly between these two areas of disjunction.

An alternative hypothesis to account for the same distributional phenomenon is that plant migration involved only short "hops" along the intervening mountains, with intermediate stations subsequently obliterated.[88] The supporting evidence is as follows. (1) Many plant species are now distributed along this mountain axis, with terminal populations in western North America and southwestern South America, and these too would become widely disjunct if the tropical segment should meet with disaster. (2) Long-distance disjunction would result in the transmission of very few alleles from the source population, whereas variability is not significantly lower in any one of the disjuncts in comparison with its homolog. Several hundred disseminules would be required to transmit sufficient genes to equalize variability. (3) Birds migrating between the disjunct areas require more than a month to make the trip. Seeds could not be carried internally for so long, and those adhering externally would have been lost by preening. Furthermore, the bird would have to start its migratory flight at the season when a plant was maturing its disseminules, and at the time of evacuation it would have to have arrived in a new area of environmental equivalence. (4) Most birds with this migration pattern are shore birds, whereas most of the disjuncts are upland plants, and often require arid climate. (5) The Parana basin of Brazil and Uruguay contains eolian sandstones which indicate at least local spots of aridity in the geologic past,[311] although fossil evidence of arid spots has not been found. Even today *Larrea* finds local spots of congenial dryness in southern Mexico, Bolivia, and Peru between the areas of more widespread representation in southwestern North America and Argentina.

As the above well illustrates, the case for long-distance dissemination is good in some situations, but controversial in others. It is usually not a "yes" or "no" possibility, but rather one of varying probability. Even if probability is very low, in the course of geologic time wide gaps may be bridged again and again. The fantastically low probability of one in a million for crossing in any one year, increases to odds of 2:1 that the crossing will be effected in the course of a million years, which is not long on the geologic time scale. Put another way if a million disseminules "attempt" the crossing in any one year, one would probably succeed.[369] Thus "impossible" is a word that must be used with caution in connection with barriers, for they tend to be relative in a time sense. It has been estimated that the 1729 known species and varieties of seed plants native to the Hawaiian Islands have been derived from about 272 original immigrations. Considering that the age of these volcanic islands is between 5 to 10 millions years, only one new colonization every 20,000–30,000 years could have produced this flora. On the other hand, unless long-distance transport is relatively rare, it is difficult to see how the differentiation of species into local races could be initiated or maintained, or why so many species have restricted ranges whereas their potential range is large.

Long-distance dissemination is another way in which disjunction and vicariism may arise that does not involve disruption of a formerly continuous range.

Polytopic origin.   Another hypothesis to account for disjunction involves the possibile independent origin of the members of a disjunct pair. It is considered possible that parallel evolution might produce the same taxonomic derivatives at widely separated points within the range of a common parental stock. For example, a widespread species with a rich gene pool might come into contact with the same highly distinctive type of habitat at different places, and at each place there could be a sorting out of those biotypes containing alleles conferring tolerance of that habitat but less successful elsewhere. In Sweden it has been suggested that the dune ecotype of *Hieraceum umbellatum* has differentiated anew at each isolated dune area, from the extensive population on intervening loams.[394] Since the dune populations are morphologically distinct, they are recognized as subspecies.

Parallel evolution is also an acceptable hypothesis where rare hybridization with the same companion species could have occurred. In North America man has created an aggressive weedy hybrid *Lonicera tartarica* ×

L. morrowi wherever he has grown these two Eurasian ornamentals in close proximity, and such might well have happened in nature.

In addition, the accidental doubling of chromosomes (autopolyploidy) might have taken place at different points over the range of a species.

Thus the possibility that disjunction could arise through parallel evolution seems a tenable hypothesis, especially at the subspecies level.

Polyphylesis.   Disjunction might arise if different parental stocks produced derivatives so closely similar that they would be considered as belonging to the same taxon. This is known as polyphylesis or convergent evolution, in contrast with parallel evolution as discussed above. However, the possibility that the same genetically complex *species* could arise from different parental stocks is so infinitesimally small that it is generally discredited. Too many genes, which give each species its character, would be involved.

On the other hand, at higher levels in the taxonomic hiearchy, where few characters bind a group together, it would be possible for a taxonomist to have drawn together in the same Order, Family, or Tribe, units that happen to share a few characters without being related by common descent. The units thus drawn into the same group would falsely appear to represent vicariads. Such faulty taxonomy has been found among grasses. For decades taxonomists had grouped grass genera into clearly differentiated Tribes believing each of these to represent a common ancestral derivation. But recently this grouping has been found less tenable than another,[158] so the old Tribes must be considered as polyphyletic.

Wherever the ranges of different genera in a family are widely separated, the possibility of taxonomic error in grouping them together should be carefully checked. If the genera prove to be of different descent the disjunction is only apparent, and no phytogeographic problem is presented. Where disjunction involves species rather than genera, polyphylesis is highly unlikely since the units would not have been included in the same taxon unless they shared many characters.

Continental drift.   We cannot be certain which if any of the extant disjunctions can be attributed to continents drifting apart after a taxon became represented on the segregating fractions. On the other hand, this seems mainly to account for ancient fossil floras which are remarkably similar but occur so far removed as India and South America. (Continental drift is discussed later.)

Hybridization

As pointed out previously, in a sexually reproducing and outcrossing species, taxonomic differentiation usually occurs only after a disjunction arises. Therefore, wherever we find the ranges of closely related species overlapping and hybridization taking place in the area of sympatry, they can be presumed to have come together following a period of separation that allowed divergence which did not progress to the point of developing incompatibility. In North America *Picea glauca* hybridizes where it makes contact with *P. rubens, P. engelmannii,* and *P. sitchensis.* Since practically all the hybrids are in glaciated territory, they are probably a consequence of postglacial range adjustments, following restriction to different areas south of the ice.

CAUSES OF MIGRATION

Intrinsic Causes: Evolution

Within a population irreversible changes in genes accidentally occur from time to time, and although most of these would have detrimental effects if they gained expression in either form or function of the individual, they are usually paired with an older dominant gene and therefore do not gain expression. Thus a hidden pool of recessive characters tends to accumulate in a population through time.

Occasionally new genes or gene combinations produce preadapted biotypes enabling a species to extend its range, as an ecotype, into an area previously beyond its limits of tolerance. The fact that some species have been shown to consist of a mosaic of ecotypes, each confined by the limits of its unique ecologic amplitude to a rather well-defined segment of the species total range, is substantial evidence that this has happened. For example, throughout its north–south extent in the midcontinental steppe of North America, *Bouteloua curtipendula* consists of populations with such distinctive genetic specialization with respect to photoperiod requirements that southerly plants cannot complete their life cycles at the latitude of the northerly races, and vice versa.[307] It is not known if this adaptation is a continuum, i.e., an ecocline, or whether it involves a chain of discrete ecotypes. In either event the past expansion of this species northward from its presumed Mexican center of origin must have been conditioned by the rate of preadaptive mutation in the peripheral populations that provided biotypes capable of enduring increasingly longer days in summer.

Range extension could also be initiated by the introduction of new genes into a species if a rare hybridization (a phenomenon known as introgression[10]) were to increase a plant's ecologic amplitude.

Extrinsic Causes: Changes in Environment

As noted earlier, when a plant has spread to occupy all of the congenial area in the existing pattern of environments, different sectors of the periphery of its range are determined by different aspects of environment. Wherever temperature or moisture are limiting, any change in climate is likely to relieve adversity along one edge of the range, while increasing adversity elsewhere. Range displacement is the common outcome, and there is abundant evidence of plants having moved up mountain slopes in response to warming, and downward in response to cooling trends. Corresponding latitudinal displacements have also been well documented.

Climatic change does not necessarily result in range expansion or displacement. Unusual gene recombinations, or recessive genes accidentally becoming paired in a zygote, may produce an individual that is better adapted to the new conditions; therefore, a new ecotype could evolve in place. If the climatic change were to procede slowly enough that adaptive evolution through successive regrouping of alleles kept pace, the range might not be altered even though the climatic regime is modified considerably, and the variety of habitats within that range is changed or reduced. But owing to the slowness of evolution in plants in relation to the rate of change in climate, migration is the more likely response. Where compensatory migration is impossible, the range may gradually diminish with the taxon ultimately facing extinction.

Before the rise of agriculture, most plants which have become weeds had small ranges and occupied niches in the nearly closed vegetation of the earth where they were probably inconspicuous. Widespread destruction or disturbance of closed communities, providing new types of environment, proved a boon to certain biotypes of these plants, and the unwitting dispersal of their seeds by man aided their spread even more by facilitating the crossing of former barriers to migration. Here the changes that initiate migration are sociologic rather than climatic.

Geologic processes can also initiate migration. Elevation or submergences of an isthmus, such as the Isthmus of Panama, can permit either intercontinental or interoceanic migration. When the strip of land is low enough that one or more segments is submerged, terrestrial species migrating toward the discontinuity tend to accumulate on either side of it, then spread across in both directions as the land is uplifted so that terrestrial environment becomes continuous. On the other hand, marine organisms are free to resume migration from one ocean to the other each time the water connection is reestablished.

DISSEMINATION EFFICIENCY

Potential Rates of Spread

Mobility of disseminules, and minimal length of life cycles, are important in determining the potential rate of migration.

Arceuthobium, a vascular partial parasite on conifers, has been observed to spread at a rate of only 0.0003 km/yr when extending its range by seeds forceably ejected from its capsules. However, birds appear to have carried the mucilage-coated seeds and extended the range of the plant at a rate of 1.6 km/yr.[182]

If a seed requires 7 hr to pass through a duck, and the bird flies at 80 km/hr, the potential rate of spread is 560 km/yr.[211]

Within a year after is introduction into the Nile River, the floating plant *Piaropus crassipes* had spread 1000 km downriver.[152]

Not only is there a tremendous variation in the potential rate of migration among plants, but prediction of this on the basis of disseminule morphology is unreliable. For example, the small winged seed of *Pinus* might seem more mobile than the heavy nut of *Corylus*, but, in fact, *Corylus* advanced across deglaciated terrain more rapidly than did *Pinus* when the last ice cap receded in Europe.[132]

Plants having closely similar ranges often have disseminules that would seem to vary widely in their efficiencies. Thus, it seems that environments change so slowly that plants with very different rates of migration have been able to keep up with the changes and maintain ranges bounded by unsuitable environment. However, the degree to which this is true undoubtedly varies from place to place. This is one of the most disconcerting unknowns in any attempt to explain the distribution of a species on the basis of existing environmental factors. In some places we are certain that the rate of migration has not been rapid enough for a species to have exploited all suitable areas. For example, since about 1900 Arctic regions have been warming to such an extent that plant ranges are generally moving northward.[164] This phenomenon is clearly manifest in the age gradients of trees; and owing to the time required for migration and the development of new forests, the evidence will be patent for many decades. Also, it is significant that many plants now confined to southeastern North America are tolerant of temperature conditions far to the north, suggesting that even in temperate latitudes migration has not kept up with recent warming. Buttressing this observation, many boreal species have recently disappeared from the most southerly stations where they were once collected. We can expect more evidence of this type to be uncovered, for at the retreating edge of migrating species, a long-lived perennial, once established, may persist for a very long time after climate has ceased to permit the establishment of its seedlings.

Mobility is probably significant in a very different connection. The more mobile a species, the more likely a mutation at one point in its range may be lost by getting swamped out.

Size of Range

Plants disseminated by spores present a paradox. Species ranges in fungi seem to average larger than those of vasculares,[37] and since viable fungus spores are as plentiful in the air in mid-Atlantic as near either North America or Europe, it would appear that direct ocean crossing is probably effective, at least from west to east, in temperate latitudes.[312] Bryophytes too show greater ubiquity than vasculares. For example, over 95% of the species found in Great Britain also occur in North America, and this is a far greater degree of floristic resemblance than one finds in vascular plants.

On the other hand, if the small size of spores were all important, the equally small-spored ferns should have relatively large ranges, whereas most of them have ranges commensurate with seed plants. And orchids have rather limited ranges despite the dustlike sizes of their seeds.

This paradox may be resolved by concluding that high mobility may favor cosmopolitanism and wide ranges, but it cannot compensate for limitations of ecologic amplitude. Perhaps the microsites favoring germination and establishment of ferns and orchids are fewer and not so widespread as those favorable to fungi and bryophytes.

Age and Area Hypothesis

After a new and successful kind of organism comes into existence, as by mutation or hybridization, its increasing population tends to expand centrifugally. Assuming infinite ecologic amplitude and suitable environment everywhere, the greater the age of this species the greater should be its range. In 1922 J. C. Willis proposed an "Age and Area" hypothesis which, reduced to its simplest form, states that the size of a species' range is proportionate to its age. Although Willis indicated briefly the possible exceptions to this logic, he dwelt on the proposition that small ranges signify recency of origin.[431]

Working with floras of tropical Africa and Asia, especially island floras, he marshaled considerable evidence which could be interpreted as supporting this hypothesis. But those floras had not been subject to strong climatic stresses, especially the relatively rapid oscillations during the Ice Ages, and this point was minimized in his presentation. Consequently, botanists grappling with phytogeographic problems in temperate and cold climates were quick to point out that many taxa with restricted ranges represent relics of formerly extensive distributions (e.g., *Sequoia*), and that barriers may have prevented many old taxa (e.g., *Eucalyptus*) from having attained wide distribution. Then too, the rapid recolonization of land vacated by the last continental glacier, resulting in the near absence of correlations between species distributions and maximal ice limits, shows how quickly plants migrate, given suitable environment. Only slow and approximately equal rates of migration would validate the Age and Area hypothesis.

In summary, the size of a range in itself is not indicative of age. Both the size and configuration of a range are usually determined primarily by barriers to migration. Most species with limited ranges actually appear to be old rather than young, and have passed the stage in which they had wider ranges.[156]

MIGRATORY ROUTES

Owing to diversity in the ecologic amplitudes of organisms, and to variations in the surface features of the earth, the spread of each species is limited to certain paths along which habitats suitable to it occur in series, with all discontinuities being narrower than dissemination range. The plant has no control over the direction its disseminules move, but with essentially random movement along all radii from the parent, some disseminules may happen to alight in a new habitat favorable to their germination and establishment, with the species possibly continuing to find new habitats equally favorable beyond. Migration may involve a somewhat linear tract of favorable environment, such as along a seacoast, a valley, or a mountain range, or it may involve a mass latitudinal or altitudinal displacement of a belt that retains much of its integrity as it shifts position at right angles to its longitudinal axis.

A feature which serves as a migratory route (or corridor, or tract) for certain species may not be favorable to others with different habitat requirements. Therefore, a route has somewhat the character of a filter in that it allows only certain types of organisms to use it. Beringia, for example, never became warm enough to allow an exchange of modern tropical genera between Asia and North America, although it proved adequate for exchanges of temperate, subarctic, and arctic genera. Again, tumbleweeds (e.g., *Salsola kali*) could not travel along a chain of closely spaced islands, as could *Compositae* with comose achenes. Owing to such sorting, ecologically related plants tend to stay together in taking advantage of a new migratory route as it becomes available.

Some specific examples of migratory routes are as follows. *Glehnia littoralis* probably originated in western North America where most of its near relatives occur, spread northward along the Pacific strand, thence westward across the Aleutian Archipelago and southward to Kamchatka.[78] Coastal migrations of this type were greatly facilitated during the Ice Ages when so much water was tied up in continental ice that sea levels dropped 100 m or more, exposing strips of land on which competition was weak.

Grayia spinosa, probably originating in the vicinity of the Mexico–United States border where most of the genus is concentrated, extended its range northward in the arid rain shadow of the Sierra Nevada and Cascade Mountains, finally reaching Washington.[108]

Silene acaulis is widely distributed in the circumpolar tundra, and the peninsulalike extension of its range southward in the Rocky Mountains to Arizona illustrates migration along the summits of these mountains. Alpine summits present the same problem to a migrant as do the islands of an archipelago, that is, they are useful only to the extent that the gaps do not exceed dissemination range. Lowered timberlines during glaciation created alpine environments on relatively low mountain summits, and this may well have been critical in enabling arctic species to penetrate so far southward.

Platanus occidentalis and *Acer saccharinum,* both trees of floodplains and river terraces, and both now occurring in southern Ontario, exemplify species that have followed river valleys as they extended their ranges northward in postglacial time.

The foregoing illustrate migratory routes along special pathways on land. In the seas water currents may have had similar influence in determining the direction of migration of strand plants.

Direction of Migration

There is usually nothing inherent in the physical character of a migratory route indicating which direction organisms may have moved along it. However, plant distribution patterns often contain evidence indicating the direction of past movements.

The single species of cactus, *Rhipsalis baccifera,* that occurs in the Old World tropics as well as tropical America, is an epiphyte with bird-disseminated seeds, and is generally interpreted as having crossed great ocean gaps, traveling eastward from its original home in the Americas where all the other species in this large family are restricted. It occurs in Africa, Madagascar, and Ceylon, and it is less likely to have reached Ceylon by having traveled westward. Thus, a single representative of a group located in a remote area may as a rule be presumed to have migrated from the place where the bulk of its relatives occur.

Where detailed maps of species' distributions are available, they often show a continuity of populations along one margin of the range, with a highly discontinuous distribution along the opposite margin (Fig. 2). This may be interpreted such that in recent geologic time the species has reached the margin where populations are continuous, and, in fact, may be in the process of advancing further there, whereas on the opposite side of the range it has been retreating, leaving stranded populations in a few habitats of compensation.

Coordinate ranges of different species can be interpreted as indicating a direction of migration if the ranges extend for differing distances from the same end of a special migratory route. Such ranges are said to be equiform. If at the staggered termini there are no disjunctions, then the plants can be thought of as having moved from the common center at different rates of speed. Such a series of ranges is said to be equiformal progressive.[194] Possibly the staggered termini reflect differences in adaptation to a climatic gradient, or to the necessity of an antecedent for the advance of some.

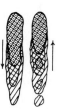

If on the other hand the ranges are discontinuous along their staggered termini, the plants have evidently vacated a formerly common territory to varying degrees as they moved into a new area favorable to all. Thus, there are equiformal retrogressive series too.[108]

Often all the plants in a given area have been classified according to the direction in which the centers of their ranges lie, and the assumption made that they have reached the given area by migration from those range centers. This assumption may be correct, but it can also be in error unless other matters are taken into consideration. For example, from the standpoint of a botanist in Minnesota, *Picea glauca, Abies balsamea,* and *Larix laricina* are species of "boreal affinity," since all have their centers of distribution well to the north of that state. But these plants attained their present distribution by entering from the south during deglaciation, and have "overshot" Minnesota. These three trees can be considered of boreal extraction only if some interglacial or much older date were taken as the point of reference. The point being made is that the present center of a range can be a treacherous indicator of the last range change. During the course of geologic time most species have migrated in different directions at different times.

BARRIERS TO MIGRATION

Once it originates, a species tends to spread in all directions until it encounters environments that prevent the completion of its life cycle. However, well beyond such barriers there are frequently other areas on which it could grow successfully. Thus, a plant's range is usually less than the total area in which it is adapted, owing to gaps between suitable habitats which exceed dissemination distance.

Population density at the margin of a range has a bearing on the effectivity of a barrier, since it governs the number of disseminules, hence, the frequency of "attempted" crossings. Adaptations for dissemination also have an obvious bearing on the efficacy of barriers. Dormancy and a durable covering over the disseminules are also helpful in crossing inhospitable territory. Finally, the broader the ecologic amplitude, the more likely a species can avail itself of habitats of compensation as stepping stones to cross a fairly impervious barrier, and the more likely its disseminules will encounter a suitable microsite in the habitat mosaic beyond.

The effectivity of a barrier is probably never exactly equal for two species, owing to differences in their types and degrees of adaptation for dissemination. A narrow strip of land would be quite effective in keeping *Nuphar polysepalum* from spreading from one pond to another nearby, but its close associate, *Typha,* would find the strip no barrier at all.

Any barrier that is less than perfect must at the same time be considered a migratory route with a high filtering effect. Also, any feature of the earth's surface that serves as a barrier to one ecologic class of organisms, usually serves as a migratory route for another ecologic class. Thus, the isthmus that connects North and South America has served as a barrier keeping the mangrove tree *Pelliciera rhizophorae* restricted to the Pacific coast, while the continuity of this narrow strip of land has allowed Andean species of mountain plants to extend north into Mexico and *Quercus* to spread south into Colombia.

Climate

Almost any facet of climate—the amount or seasonal distribution of precipitation, amount of summer heat, occurrence of frost, abundance of fog, etc.—can be critical for a plant, depending on its special environmental requirements and tolerances.[200] Examples of climatic restrictions on plant migration are therefore numerous.

At the Canada–United States border many species of plants are abundant on those slopes of major Rocky Mountain ranges which face into westerly winds and, therefore, have a somewhat oceanic climate. These same species are poorly represented if not completely absent on the lee slopes of the same ranges, as well as on all slopes of these mountains to the north and south of the border where oceanic influence is feeble or lacking. *Pinus contorta* has migrated up the slopes of the Rockies to elevations near upper timberline, but the *Arceuthobium* that parasitizes this pine finds its climatic limits lower on these slopes. African species of mangrove are different on the eastern and western coasts since the temperate climate of the Cape of Good Hope prevents these frost-sensitive plants from migrating around the southern tip of the continent.

Commonly, by taking advantage of the most suitable microenvironments among those available, a plant can penetrate far into a region where the climate is relatively unfavorable. Thus, *Juglans nigra* and *Carya cordiformis*, mesophytic trees widely distributed on uplands in the temperate forests of eastern United States have linear extensions projecting far westward into the midcontinental steppe, owing to the continuous moisture supply and protection from desiccating winds that are available in ravine bottoms. Climate becomes a completely effective barrier only when *all* habitats lie outside the ecologic amplitude of a species.

Mountain chains may act as barriers when all the passes rise so high above the basal plains that climate prevents the crossing of plains species. But mountains nearly always separate distinctive climatic provinces, so it is not easy to determine if it is the cold montane climate, or the difference in climates on the two basal plains, that accounts for floristic differences in lowland floras.

Climatic barriers to migration are the most dynamic of all, for after only a few years with abnormally favorable weather, pronounced range extension may be evident.[146,418] It is perfectly logical to hypothesize that distributional limits are set by occasional episodes of climatic stress, but this hypothesis is singularly lacking in factual support over most of the earth's surface. Frost damage to the foliage of *Sequoia sempervirens*[269] or other trees during an extreme winter,[141] or even the killing of all aerial shoots of a woody plant,[101] tell us nothing about the climatic factors setting the margins of these species' ranges. Only when there is a complete kill of all individuals at the margin of a range can we conclude that an extreme climatic condition sets range limits. At range limits the seedlings of sprouting species may get established in favorable seasons, then the established plant may survive many subsequent episodes of weather that kill it back to the ground.

Hydrophytes exhibit considerable indifference to climate, in comparison with mesophytes and xerophytes. There is so little difference among marsh and pond habitats that species such as *Elodea canadensis, Phragmites communis, Potamogeton pectinatus,* and *Typha latifolia* range over a wide variety of climates.

In oceans the boundaries between masses of water of different temperatures often serves as barriers comparable to climatic limitations on land.

Oceans and Seas

There are quite a few species of pantropical plants, but no pantropical species of mammals. Therefore, it is clear that mammalian distribution is conditioned primarily by land connections and only secondarily by climate, whereas the reverse is true for terrestrial vasculares. Nevertheless, despite a number of interesting similarities among the floras of tropical America, Africa, and Asia, there are more differences in this latitudinal belt than exist between the continents in the north temperate and especially arctic belts. At high latitudes there have been land connections at times in the past that have allowed floristic exchange, but in tropical latitudes the oceans have been major barriers practically throughout the history of angiosperms, except for plants that are adapted to float long distances in salt water, and those that were able to take advantage of temporarily available volcanic islands which served as stepping stones. Even among strand species that have floating disseminules, the existence of geographically distinct subspecies shows that there is a degree of isolation even if it is not very great. The distinctiveness of mangrove floras on the east and west margins of the Pacific Ocean, and especially their absence in the native flora of Hawaii, shows that oceans can be barriers to even water-disseminated strand plants if the oceans are sufficiently vast.

The relationship between the flora of Madagascar and adjacent Africa provides an instructive example of the effectivity of salt water as a barrier to plant migration. Although the strait is only about 400 km wide, the two floras are sharply distinct, with more than half the Madagascar flora being endemic. This circumstance is especially significant with respect to the concept of stepping stones as intercontinental migratory routes. Islands must obviously be closer than 400 km for effective exchange. At the present time there are no connections this close to allow transoceanic migration in the southern hemisphere, except among the islands south of Asia. In the north the distances between Greenland, Iceland, Faraoes, Shetland, and Scotland all slightly exceed the distance between Madagascar and Africa. Only the Bering Strait, 83 km wide, would seem amenable to intercontinental plant dissemination at high latitudes at present.

Direct evidence of transoceanic transport of viable disseminules has been reported by biologists interested in the revegetation of the completely glaciated Macquarie Island that lies 30 km from the next nearest land. A pelagic bird was observed to carry a viable seed of a land plant to this island. Also a viable seed of *Caesalpinia bundoc* was found along the shore, and it very likely had floated some 20,000 km from Central America, remaining in salt water for at least a year.[81]

Land Masses

In its broader aspects, the distribution of marine organisms is strongly determined by the pole–equator gradient in ocean temperature. Each of these organisms tends to extend its range around the earth within a particular temperature belt, wherever it can find depths and other conditions matching its requirements. North–south oriented land masses, such as North–Central–South America and Eurasia–Africa, form effective barriers for marine organisms intolerant of very low water temperatures near the poles, even where a continuity of land is maintained by only a narrow isthmus. As an illustration, six species of Indo–Pacific algae have spread from the Red Sea to the Mediterranean since the Suez Canal was dug,[5] despite appreciable temperature differences between the two water bodies, and the high salinity of the Bitter Lakes between segments of the Canal. In the Western Hemisphere fresh water in the Panama Canal has supplanted land as a barrier separating Atlantic and Pacific marine biotas, but the tentative plans for a sea-level canal paralleling the Panama Canal has elicited concern among marine biologists for the possible consequences of extensive comingling of the two ocean biotas. Strong competition following free intermigration is predicted to allow strong competitors to bring about extinctions, thus reducing biotic diversity of the area as a whole.[50]

Soils

Certain plants have such special soil requirements that large tracts of common soil types prevent their spread. For example, serpentine-requiring plants[273] have extremely little opportunity to migrate from one serpentine outcrop to another because these areas are widely separated by other kinds of soil. Also, plants requiring sand often have been unable to reach remote tracts of suitable substrate.

East of Hudson's Bay vascular plants requiring nothing more special than a rooting medium of loose mineral material are restricted in at least their local distribution as a result of glaciation having scoured away nearly all loose material down to bedrock.[276]

Daylength

An interbreeding plant population often tends to become genetically attuned to the annual cycle of daylength variation at any latitude where it persists for a long time. Responses to daylength then serve as a timing mechanism keeping the annual cycle of phenologic phases synchronized with appropriate segments of the annual cycle of climate. It prevents too early bud-break at the end of a cold season, and too late a development of dormancy at the approach of the cold season, either of which might result from abnormally warm weather. Adjustments to a dry season can also be controlled in this way. But such a development of genetic dependence upon the photoperiod becomes a threat to survival if climatic change should necessitate migration either northward or southward. Thus, the photoperiod can serve as a barrier to migration, either when movement is a requirement, or when range extension becomes possible by the elimination of some other limiting factor. However, it is doubtful if this is a very important restriction on migration, for it appears that environmental modification requiring migration is normally slow in relation to the rate at which plants can adjust to a different photoperiod. During the Ice Ages many plants shifted their ranges great distances southward and then northward alternately. Presumably rich pools of alleles allowed new ecotypes to arise successively along the advancing front of a range. At the same time, there may well have been biotype depletion along the receding margin. It is possible that some taxa present before the first of this series of glacial episodes lacked sufficient genetic diversity to adapt and so became extinct.

Photoperiodism is undoubtedly crucial with regard to long-distance dispersal in a north–south direction. It is probably significant that those plants which are restricted to the high mountains of tropical Africa, and which have been derived from temperate zone ancestors, are all day-neutral. Only this character would preadapt a plant of the temperate zone for success on a tropical mountain after long-distance dissemination.[297]

Range of Obligate Symbionts

The completion of the life cycle of many organisms is absolutely dependent on some service or substance that can be supplied by only one or at most a few other organisms. Any condition that acts as a barrier to an essential symbiont also serves indirectly as a barrier to the organisms dependent upon it. *Epifagus*, a vascular root parasite on *Fagus*, was able to migrate from North America to tropical Mexico only as its host migrated in past geologic ages. *Pinus radiata* prospered as an introduced forest tree in Australia only after the fungus needed to form mycorrhizae with its roots was also brought to that continent. Smyrna figs remained unproductive in California until the essential pollinating wasp was likewise imported from the Old World. The absence of a suitable disseminating agent could likewise be a barrier to migration.

Vegetation

Nearly all plants can be grown outside their natural ranges if competition is kept to a low level. The absolute confinement of many introduced weeds to ground where disturbance has thinned the plant cover provides substantial evidence that closed vegetation limits the spread of many species. Past events that created thinly populated land areas must, therefore, have been the impetus for many range extensions. For example, when continental glaciation locked up so much water as ice on land and sea levels were lowered, exposed strips of coastline must have greatly facilitated strand plants in crossing barriers to reach new areas of permanently favorable environment. Also, islands emerging from a sea or emerging after having been covered by ice, would be much more hospitable to immigrants in the early stages of colonization. Seral species that prosper on bare mineral soil would be most favored at first, with the balance later shifting to favor shade plants and good cometitors.

Floristic and faunistic biogeography have much in common despite some differences in principles. Vegetation areas set limits to the distribution of both plant and animal species. Also, animal ranges, like those of plants, rarely show much correlation with a particular community type.

SOME CONSEQUENCES OF MIGRATION

Evolutionary Change

As a range becomes large, distance begins to impede gene flow, so that populations remote from each other may start to evolve different characteristics.[187]

Enforced migration may also effect evolutionary change by eliminating certain ecotypes, while at the same time providing opportunities for the development of new ecotypes along the advancing margin, either with or without introgression. Concurrently, the disjunctions that develop or become more absolute along the receding edge tend to initiate evolutionary divergence among the relictual populations.

The first species to reach a new oceanic island meet little competition, so that genetic innovations have a high chance of survival, at least if the immigrants are heliophytes or facultative heliophytes. Eventually, as populations thicken, variations in shade become added to variations in soil and topography, providing a wider variety of microsites inviting invasion. Evolutionary divergence is a common consequence of migration, especially where a new land area is involved.

Environmental Restriction

A progressive reduction of ecologic amplitude commonly results as a migrating plant approaches its climatic limits. Fewer and fewer habitats meet the requirements for germination and completing the life cycle. Thus, along their relatively cold high-altitude limits in the Rockies, *Quercus gambelii, Pinus ponderosa,* and *Pseudotsuga menziesii* become restricted to warm south-facing slopes, or to rock outcrops where a closed vegetation cannot produce cool microclimates with its shade. At the same altitudes, cool north-facing slopes, ravine bottoms, and frost pockets are the only habitats that harbor outposts of trees characteristic of higher altitudes, such as *Picea engelmannii* and *Abies lasiocarpa,* which reach their lowest limits there.[98]

Autecologic Maladjustments

As a species reaches its climatically determine range limit, the form and function of marginal individuals may be strongly altered. Vegetative parts are commonly dwarfed as trees approach cold timberlines or arid regions.[97] At these limits seeds may be produced only at rare intervals.[207] Some species change from their normal bisexual to unisexual habits, and still others may be limited to only vegetative reproduction.[133]

But maladjustments such as these are by no means the rule. Should the critical limitation be simply a matter of seedling survival, the life cycle of those marginal individuals that get established in exceptionally favorable microsites or during exceptionally favorable weather, may then be completed without evidence of stress. In northern Idaho the size and productivity of *Agropyron spicatum* are greatest at its uppermost limits in the mountains.

Polyphyletic Floras

Paleobotany indicates long histories for at least the majority of floras. With the passing of time they have expanded and contracted repeatedly, occasionally incorporating new floristic elements by immigration or evolution, and from time to time suffering the loss of members. The present composition of the flora in a limited area, therefore, represents no more than a cross section of changing ranges that appear fixed only because the brevity of man's life span strongly conditions his perspective.

So complete have been the past changes in plant distribution, that with the possible exception of certain parts of the moist tropics the floras of every portion of the land surface appear to have been wholly replaced time and again since angiosperms appeared. Evidence of such complex histories may be found in nearly every area if the species are grouped according to their present distribution patterns, as these suggest their most probable migratory histories, i.e., sources and migratory routes. Several types of floristic affinities are usually evident, and it is often seen that each floristic element is at present characteristic of some special segment of the environmental mosaic, and related to a distinctive migratory route. Furthermore, it is rather unlikely that the historic event which brought one ecologic group into the area could have brought in all the others as well, so that by inference at least, some of the diverse floristic elements are of different ages. Thus the floras of most regions are polyphyletic, built up of accretions differing in ecologic character that came as successive waves of immigration from different directions, and tending to concentrate in different habitat types. Also, any climatic change allowing immigration from one direction is likely to have forced simultaneous emigrations in the opposite direction, leaving relics of opposite affinity.

The plant life of the Bitterroot Mountains and adjacent plain to the west provides a good example of a polyphyletic flora.[108] At middle altitudes there is a temperate mesophytic flora that invaded the area from the northeast or descended from mountain summits early in Cenozoic time, as climatic cooling caused a semitropical flora to move southward. Later a mesophytic subalpine and alpine flora came southward along the summits of the Rocky Mountains as they uplifted and became too cold for most temperate plants. At a still later date xerophytes invaded the region from the south when the Cascade Mountains near the Pacific Coast became high enough to produce a rain shadow along their eastern flank. Thus in ascending these slopes one crosses three major floristic belts, each differing as to its antiquity, origin, and its present position on the climatic gradient.

Not only is a flora as a whole polyphyletic, but the same plant association usually includes species that have come from different source areas at different times, as can again be shown by floristic analysis. Since the genetic character of no two species is alike, the members of an association cannot respond in exactly the same way to a given change in climate. Hence, as a climate changes some species may leave the association and new ones enter. This is not to deny that many species in a community have sufficient similarity in their ecologic requirements that the community can shift its range and retain its identity despite some changes in its composition.

ENDEMISM AND ITS INTERPRETATION

The adjective cosmopolitan implies that a taxon occurs throughout the world. Possibly some bacteria may be cosmopolitan, but certainly no higher organism has sufficient ecologic amplitude to be cosmopolitan in the sense of having unbroken gene flow throughout. The closest approach to cosmopolitanism in vasculares seems to be the few species that occur on all continents, such as *Phragmites communis* and *Pteridium aquilinum.*

The term endemic is used when indicating the restriction of the range of a taxon to a single geographic unit, although the degree of restriction varies widely. *Picea glauca* may be said to be endemic to North America. *Sequoia* is endemic to coastal California and Oregon. *Dasynotus* is endemic to a few square kilometers in northern Idaho.

Endemism, especially narrow endemism, immediately arouses interest in the cause of the restriction, and sufficient evidence has been brought forth to substantiate four types of endemism.

Residue of Decimation: Paleoendemics

Wallace's model of the normal history of a taxon implies that plants with small ranges may represent remnants of formerly widespread distributions. As for *Sequoia*, paleontology often provides proof of a much more extensive ancestral range, followed by catastrophic elimination of much of the suitable environment by climatic changes, continental submergence, etc. Such relics are paleoendemics.

As details of plant distribution in Scandinavia were worked out during the 19th century, it became apparent that certain species are endemic to either or both of two limited areas in Norway: a northerly strip along the Atlantic Coast, and a southerly area in the mountains back from the coast. Twenty-five species of the Norwegian flora were found to have a bicentric distribution, with 40 additional species restricted to one or the other of these two areas. Interest in these endemics was greatly increased by the discovery that they are accompanied by a still larger group of disjuncts that occur in other parts of the world but in Scandinavia are found only in these places. In 1876 Axel Blytt hypothesized that these islands of endemism and disjunction represent places not covered by the last continental glaciation (now called nunataks), on which the plants survived the cold period, and from which they have subsequently shown little tendency to spread.[154] Several types of physical evidence support this interpretation. (1) When the normal slope of the ice surface at the margin of a major glacier is projected inland from the end moraine of the last glacier, the peaks of the southern nunatak area are left projecting above the probable ice surface. (2) The refugia provide an abundance of steep equatorward slopes which provide abnormally warm microclimates. (3) The primary minerals of residual soils on the hypothecated unglaciated land are more weathered than would be expected if pedogenesis had not started until after the last glaciation, and these soils are in exposed positions where an ice cover would have scoured them away.

Botanical support is also of several types. (1) The number of species having such a peculiar distribution would be hard to explain on any climatic, edaphic, or biotic basis. (2) Pollens preserved in bogs in the vicinity of the nunataks show that the species restricted to the nunataks cast pollen into the ponds of the immediate area well before these types of pollen appear in the record of areas remote from the nunatak. (3) In the southern refugium *Papaver radicatum* consists of six races with small, mutually exclusive ranges. These races differ as much from each other as they do from other races of the same species that occur in the northern refugium, in the Faeroes, and in Iceland, and this genetic differentiation could hardly have developed since glaciation. Therefore, all must be of approximately equal age, with intervening populations having been eliminated during glaciation. The southern refugium is well back from the coast, and the species restricted to it are intolerant of maritime climate, and so could not have moved inland from the coast in postglacial time. (4) Plants growing on existing nunataks elsewhere prove that many plants can maintain populations on small knobs of land projecting above the surface of a glacier.

It is an interesting point that geologists have been more than reluctant to accept biologic evidence of this sort as indicative of areas free of ice during the last glaciation. Possibly some of the differences between botanists and geologists can be resolved as a result of the discovery that the seeds of some plants can remain viable for a long time at near-freezing or lower temperatures, so that soil only briefly covered could have retained viable seeds.

In the early part of the present century M. L. Fernald made extensive plant collections in southeastern Canada, and found that a group of disjuncts and endemics are centered on the Gaspé Peninsula just south of the mouth of the St. Lawrence River. He interpreted the low mountains there as nunataks that accounted for the rarity of these plants.[136] To account for their failure to spread from this limited area despite the opportunity presented by some 11,000 years of ice-free environment, Fernald hypothesized that the severe stresses of glaciation allowed only a few biotypes to find suitable habitats to compensate for the climatic adversity, so that later when climates improved the survivors did not have sufficient versatility to expand into the rich mosaic of new habitats as they appeared. In addition to this development of disharmony between the survivors and the new environments, the species would have needed to possess sufficient genetic diversity that selection could shift progressively and so provide new ecotypes adapted to the still different environments encountered during range expansion.

For a time paleoendemics were referred to as "senescent," but upon reflection this seemed hardly appropriate, for rather than continuing to decline, there would seem to be no reason why variability could not accumulate slowly and allow future spread. Furthermore, many narrow endemics can be easily grown in environments very different from those of their natural ranges; this demonstrating that they still retain considerable ecologic amplitude. Also, *Chamaecyparis lawsoniana*, with a very small range in southwestern Oregon, has a remarkable ecologic amplitude, occurring from sea level to high altitude in the mountains, from dune to normal loams to serpentine outcrops! Certainly narrow ecologic amplitude is not necessarily involved in the interpretation of paleoendemics.

Fernald's work stimulated much interest in nunatak survival in North America, and small gene pools seem quite acceptable as a possible explanation of at least certain nunatak endemics, but some of Fernald's basic data have been found faulty.[138,346,347,440] The phytogeographic work in Norway stands out as the classic study of nunatak survival.

Another example of paleoendemism in North America has claimed considerable attention. Owing to the erratic direction of the movement of ice lobes at the irregular margin of the last continental glacier, a small area in southwestern Wisconsin and adjacent parts of Minnesota and Iowa was encircled by ice and accidentally escaped glaciation. Several disjuncts and one endemic are confined, or nearly confined, to the unglaciated area which served as a refugium.[129] *Erythronium propullans*, the only endemic species, now occurs in Minnesota adjacent to the unglaciated area, and could have simply undergone range displacement. *Rhododendron lapponicum*, a widespread arctic species, has a highly disjunct population in this unglaciated area. *Chrysosplenium ioense* and *Montia chamissonis* are disjuncts better represented in the Rocky Mountains. *Sullivantia renifolia* and *Pellaea atropurpurea* are disjuncts with wide distribution well to the south of this unglaciated area.

Several additional species with limited ranges outside the refugium, but somewhat centered on it, are also probably paleoendemics that have been better able to spread: *Asclepias meadii*, *Lespedeza leptostachya*, and *Talinum rugospermum*. In a refugium as ecologically varied as this one, which provides cliffs facing all directions, genetic diversity may have been preserved in many species, enabling them to spread sufficiently to obscure evidence of their persistence there during the period of stress.

Species of limited ranges along the coasts of the North Atlantic, North Pacific, and Arctic Oceans suggest that there were a number of small coastal refugia permitting survival, despite geologic opinion to the contrary. Evidence that caribou survived glaciation on Greenland automatically implies that considerable vegetation survived there too.

Product of Long Isolation: Insular Endemics

The role of isolation permitting endemism to arise through evolution is well illustrated by insular floras, for certain islands provide the best examples of areas that have long been surrounded by nearly impassable barriers.

As A. R. Wallace pointed out in his classic book "Island Life" (1880), islands are of two general types. Some are hills arising above shallow seas that cover continental shelves (e.g., Cuba, British Isles, Philippines) and in the geologic sense are only temporarily separated from the remainder of the continental land mass, for through time all parts of a continent are subject to alternate submergence and emergence. The floras of these islands are relatively rich and have affinities with the nearby mainland in proportion to the time and distance of separation. Families well represented on the mainland are for the most part represented on the islands too.

Other islands represent the summits of volcanic peaks that have been built up from the ocean floor (e.g., Hawaii, Bermuda, Ascension Island). From their inception these are separated from land masses until they erode away or become submerged by some other process. If situated far from any continent, these are the islands that are biologically most remarkable. Their biotas originate through chance dissemination over long distances by means of floating disseminules, by rafting, by wind, or by birds.[63] *Ipomoea pescaprae* and *Cocos nucifera* are strand plants widely distributed in tropical latitudes. They have disseminules capable of floating for long periods in salt water and then producing seedlings adapted to saline sands of wave-beaten coastlines. The abundance of wood-boring insects on oceanic islands lends strong support to a hypothesis that some viable seeds may also have crossed salt water by rafting. Seeds entrapped in small pockets of soil surrounded by roots of trees could be carried long distances by a floating log, then be cast upon a shore where the log could rot. Floating masses of land vegetation have been observed 160 km out from the mouth of the Ganges River. Spiders, many of which spin parachutelike webs, are nearly always well represented on oceanic islands, and plants with comose or minute seeds would seem equally as well adapted to travelling long distances in violent wind storms. Viable seeds of land plants have been found on the feet and beaks and in the regurgitate of pelagic birds that visited Macquarie Island, an island far to the south of Australia that has been slowly regaining a flora since glaciation destroyed all previous plant life.[388]

Even with all these possibilities, few land species other than strand plants possess the ecologic requirements for even a rare accidental crossing of so formidable a barrier as broad expanses of salt water. But once a successful crossing is made to a thinly populated island, competitive abilities have little survival value and can be lost in genetic drift. At the same time new characters are not so likely to get swamped out in a large population, and so can become fixed by inbreeding. These new characters are mainly neutral rather than adaptive, as illustrated by *Pritchardia,* a palm genus which on the Hawaiian Islands has developed some 30 different species.[371] Evolution progresses rapidly, so that the affinities of the initial immigrants become progressively more obscure. Thus, in the Hawaiian Islands, which appear to have emerged in Pliocene time, about 90% of the species and 20% of the genera of indigenous flowering plants are endemic, having gained their individuality since immigration.

Another factor that is probably important in the evolution of insular endemics is that islands surrounded by temperature-stabilizing bodies of water are not subject to wide seasonal fluctuations in either climate or weather, such as repeatedly purge land floras of their accumulated variation.

The older the island and the more complete its separation from other land, the higher the percentage of its endemics and the more uncertain we are of their sources. Since the Hawaiian endemics for the most part seem to represent immigrations long past, it appears that the chain of islands that were most instrumental in their population was broken long ago.

In addition to including relatively few genera, the floras of oceanic islands are peculiar in that interspecific balances develop that are attuned to only the local biota, and when exotics with undiminished competitive abilities are introduced, the insular species are relatively incapable of meeting the new biotic pressures. Hawaii and New Zealand are outstanding examples of the disastrous impact that alien plants and animals have exerted on indigenes since the development of commerce.

Still another important character of oceanic island biotas should be noted. If there were equal probabilities of all taxonomic groups migrating to an oceanic island, the groups arriving there would be a strictly random selection of species from among the floras of neighboring continents. Large families would be represented in proportion to their sizes. However, on these islands many genera and families that might be expected on the basis of their representation in source areas are usually wholly lacking, so that the floras are said to be "unbalanced" or "disharmonic." Land plants with heavy and nonfloating disseminules, as well as mammals and amphibians, are conspicuously underrepresented on oceanic islands. The consistency of this among island biotas argues strongly against an alternative hypothesis that the imbalance is a result of random extinctions. Imbalance becomes entrenched as an island becomes covered with vegetation in consequence of the rise of at least weak competition, which can be unfavorable to latecomers arriving in a weakened condition.

On continents, those mountains which have long been isolated from other mountains usually support floras on their summits that are homologous with the floras of oceanic islands. Inimical climates on the lowlands are fully as effective in preventing the mixing of montane floras, as oceans are for insular floras. Thus about 80% of the alpine plant species of the isolated mountains of equatorial Africa are endemic, which compares favorably with the figure for the Hawaiian Islands.[185]

The size of an oceanic island tends to be positively correlated with its biotic diversity, for the larger the island the greater the variety of habitats it has to offer to immigrants and mutants, and the larger the "target" area. An oceanic island consisting of only a small coral reef may have no endemics, its flora consisting of only strand plants well adapted for dissemination by ocean currents.

The floras of oceanic islands are generally accepted as providing evidence of successful long-distance dissemination, but at the same time they point up the rarity of such an event.

Product of Recent Evolution: Neoendemics

Willis based his "Age and Area" hypothesis upon his intensive studies of plant distribution in the tropics, where endemics of relic status seemed uncommon. Although his interpretation of the many endemics that do occur in the tropics as products of recent evolution is probably correct, the phenomenon is by no means confined to the tropics. A number of temperate zone genera have an abundance of weakly differentiated endemics that are of undoubtedly recent origin, the genus *Crataegus* in southeastern North America providing an especially interesting example.

These small trees are intolerant of shade, and before the advent of white man there were probably few species, represented by few individuals, scattered widely over the nearly continuous forest. It is theorized that extensive clearing of the forests, with a consequent creation of abundant forest-margin, fence-row, wild-pasture, and abandoned-field habitats to which these thorny bird-disseminated plants are eminently adapted, permitted the original taxa to spread rapidly. Where formerly isolated populations came into contact hybrids formed, giving rise to the present taxonomically confused situation. Over 1100 species and varieties have been described, some known only from a single tree, and most with very limited ranges! The fact that the same taxon is never found in widely separated places argues strongly for recency of origin.

Rubus presents a parallel example. Three competent taxonomists, monographing essentially the same area, the northeastern United States, recognized 24, 205, and 381 species in this genus![61]

The phenomenon illustrated by *Crataegus* and *Rubus* is taking place in other taxonomic groups as well, but mostly on a much smaller scale and at a much slower rate. Man's disturbance of the environment (including the introduction of the honeybee into New Zealand!) has catalyzed evolution by hybridization, thus leading to neoendemism. But new conditions cannot have so pronounced an effect except in those species that are prone to instability at the start.

The difference between neoendemism and paleoendemism is linked to differences in the rates of evolution among organisms. Neoendemism is favored by rapid evolution, whereas paleoendemism is favored by evolution that is slow in relation to changes in the configuration of the earth's surface which disrupt ranges and barriers. Since, in general, plant evolution is relatively slow, a hypothesis of paleoendemism is generally favored over neoendemism, and at the generic level fossils often settle the status. On the other hand, where a widespread and likely parent has persisted in the vicinity of a slenderly differentiated endemic, or where disturbed landscapes have allowed members of different communities to intermingle, the probability of neoendemism increases. If a polyploid has a limited range in comparison with its diploid parent neoendemism is indicated, but if this situation is reversed the absolute age of the pair is in doubt even though the polyploid is still the derived member.

It is remarkable that whereas evolution has been so rapid in *Crataegus* and *Rubus* such as described above, other plants like *Ginkgo biloba* and *Metasequoia glyptostroboides,* are essentially "living fossils" which appear to have remained stable for millions of years.

Habitat Specialization: Ecologic Endemics

Still another cause of endemism is the necessary confinement of plants with special requirements to the limited areas that meet these requirements. Examples are provided by *Cupressus sargentii*, confined to serpentine soils of California and Oregon, and *Fouquieria shrevei* confined to gypsum soils in northern Mexico. Restrictions of this nature most likely arise through the rare concomitance of a mutation or hybridization producing an individual preadapted to an environment different from that of the parent stock(s), and the fortuitous proximity of that special environment. The geologic age of the special environment suggests a maximum age for the endemics it supports, but if possible ancestral relatives still occur in different habitats nearby, the ecologic endemics may be much younger than the special environment.

Transplant experiments have uncovered an enormous amount of ecologic endemism at the ecotype level, which passes unnoticed simply because most of these units cannot be distinguished morphologically.

Fate of Endemics

A number of paleoendemics that were crowded near the point of extinction during the Ice Ages have taken advantage of subsequently less rigorous climates and today have a far lower degree of endemism. The ranges of many of these taxa are undoubtedly still expanding. On the other hand, the changes which brought about a severe restriction of the ranges of certain paleoendemics may still be in progress, so that some of these plants are headed toward extinction. The complete disappearance of *Franklinia alatamaha*, a beautiful tree found but once in Georgia then never seen again, or the complete disappearance of *Ginkgo biloba* from natural vegetation, both illustrate such a fate. The current trituration of landscapes by massive engineering projects has made biologists apprehensive over the fate of many "rare and endangered species" of both plants and animals in the near future.

The conditions that long prevented the spread of insular endemics may weaken, and such plants, like neoendemics, could become less narrowly endemic with the passing of time.

INTERRELATIONS AMONG FLORISTIC PLANT GEOGRAPHY, TAXONOMY, AND GEOLOGY

Historical geology has demonstrated that the pattern of mountains and plains, lands and seas, has changed from time to time. Such changes have determined the options for the spread of organisms as they evolved. The changes may have created new areas of favorable environment in places too remote for an organism to colonize, or they may have broken up a formerly continuous range. Thus, the history of the earth's surface as worked out by geologists must be in accord with the distribution of modern taxa and their genetic relationships as worked out by biologists. Any discord indicates that a mistake has been made somewhere in interpretation. The intimacy of this

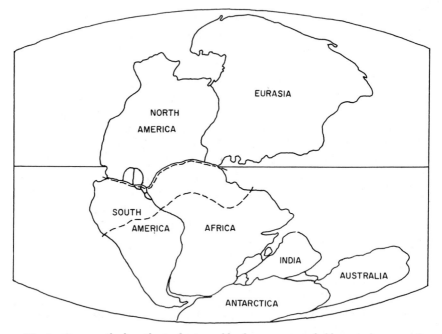

Fig. 3.   Pangaea, the hypothesized primeval land mass, surrounded by a single ocean (after Dietz and Holden[115]). The broken lines indicate possible positions of the first division of Pangaea into northern Laurasia and the southern Gondwanaland.

interrelationship is well illustrated by the theory of continental drift, which has come to be accepted by practically all scientists despite some difficulties with the theory which have yet to be resolved.[170,202,205,214,282]

In 1620 Francis Bacon became impressed by the accordant outlines of the east and west coasts of the Atlantic Ocean, as evident on world maps that were becoming increasingly more accurate, and opined that these land masses were once joined. Although others also participated in the same speculation, it was not until 1912 that Alfred Wegener assembled enough data to show that the hypothesis merited serious consideration. Wegener conceived of a single primeval continent which he named Pangaea (Fig. 3), and which subsequently divided, with the fragments drifting apart to become the continents as we know them. In 1937 Alex L. du Toit pointed out that well before angiosperms had clothed the land, any such primeval continent had divided into two parts, a southern one which he called Gondwanaland consisting of the present Antarctica, Australia, India, Madagascar, and parts of Africa and South America, all of which had been subject to glaciation between the Devonian and Permian Periods. The sister land mass, Laurasia, could not have been near a pole since it bore tropical vegetation at that time. The distinctiveness of these two continents, as well as the unity of

Gondwanaland is proven by the distribution of fossils, for a distinctive group of Permian genera called the *Glossopteris* (a seed fern) Flora is common to all parts of the former Gondwanaland, yet unknown elsewhere. Conifers did not develop much until after the first division of Pangaea, for modern genera with few exceptions are remarkably restricted to either the North or South Hemisphere, yet are widely distributed in their own hemisphere, e.g., *Dacrydium* and *Podocarpus* in the South Hemisphere, versus *Abies* and *Picea* in the North Hemisphere. Appreciable floristic resemblances among the modern angiosperm floras of South America, Africa, and Australia indicate that the fragments of Gondwanaland did not become completely separated until after a number of modern families had evolved and become widespread.

Just when did the drifting apart occur? The ubiquity of fossil dinosaur genera plus the pan-tropical distribution of all higher taxa of modern reptiles implies that, although separate continents probably had existed prior to Pangaea (Fig. 4), continental connections existed after reptiles evolved and became widely distributed in Cretaceous time. Yet by the end of Cretaceous time all the present oceanic barriers had become effective at least for mammals, for the distribution pattern of mammals which began to evolve rapidly in that Period indicates that their spread has been consistently limited by the present pattern of land and ocean. Since both vascular plants and reptiles

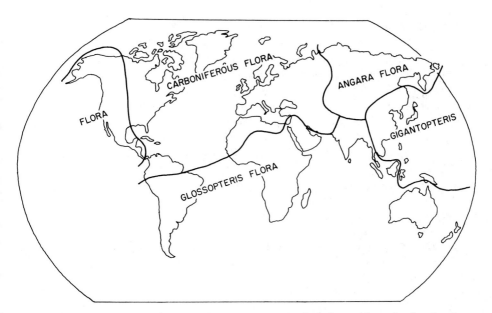

Fig. 4. Paleozoic–early Mesozoic floral regions, each of which must have developed and attained its range on a distinctive yet continuous land mass.[286]

were well along in their evolution before the last major segregation of land masses, whereas mammalian evolution has mostly postdated that event, the present continental segregation is rather well fixed as a Cretaceous phenomenon. Other segregations, and possibly reunions, had occurred previously.

Floristic plant geography must look upon continental drift as an agent of evolution as well as of dispersal, for once the land area became discontinuous evolution was accelerated. Pangaea was such a large land mass that the climate at the center could have been extremely continental, and certainly the absence of ocean barriers to dispersal would have dampened evolutionary progress.[368] Later, as the land mass became divided into progressively smaller units, this must have favored evolution not only by the barrier effect but by increasing the area of oceanic climates. Where change in latitude was involved, extinctions must have eliminated some elements while favoring preadapted subordinates and their evolutionary derivatives. Thus, the diversity of the earth's biota would have increased. Then when fragments that had become segregated for a time came together (as did the northern and southern parts of Africa and South America), the proportion of oceanic climates decreased, with continentality increasing in the interior. Superior competitors in the pooled biotas would alter competitive balances eliminating inferior competitors and favoring evolution by hybridization as well. If collision produced a mountain ridge at the juncture (e.g., the Himalayas), a new barrier would arise after a brief period of biotic intermingling. Certain pantropical species, e.g., *Gyrocarpus americanus*, may have gained their widespread representation on Pangaea before the Cretaceous disruption was complete, then retained their genetic character through the following eons, indicating that major displacements could not have involved much shifting to high latitudes, otherwise this tropical species would have lost out. Some of the evidence interpreted as indicating mass migrations, or as indicating past barriers will need reinterpretation in the light of the recent deepening of our knowledge of continental drift phenomena.[332]

Currently the earth's crust is conceived as a mosaic of half a dozen or more rigid plates, each including both continents and ocean basin topography. These plates slowly move about in relation to one another. Along some of the contacts lava upwells to produce ridges, especially along midocean lines. Elsewhere new lava forces plates apart to create ocean trenches and in so doing alters intercontinental distances. At some contacts one plate rides over another forcing it down to once more become part of the central molten core.

Floristic evidence indicates that the gaps between continents arose one by one, well spread through the period when angiosperms have been evolving. For example, both the *Proteaceae* and *Restionaceae* show that closer relationships exist between Australia and Africa than exist between either and South America. This suggests that Africa and South America separated before Africa lost its contact with Australia by way of Antarctica.[91] Investigations of the past and present movements of these plates forming the earth's crust is a field currently quite fluid, and its impact on current concepts in geography will probably be great.

Aside from the direct significance of continental drift for plant geography, the topic shows clearly that there are indissoluble ties among physical geology, paleontology, plant geography, and taxonomy. Firm evidence in one field imposes restraints on theories in the other fields, but at the same time suggests new hypotheses for testing.

Part II

ECOLOGIC PLANT GEOGRAPHY

INTRODUCTION

In 1805 Alexander von Humboldt, in his "Essay on the Geography of Plants," pointed out that widely separated places over the earth have physiognomically similar vegetation if their climates are similar, and in 1817 he pioneered in attempting to explain vegetation distribution on an environmental basis, with emphasis on temperature. The importance of the moisture aspect of climate was slow to gain recognition, and before moisture and temperature relations were reasonably well integrated into a concept of climate, V. V. Dokuchayev and other Russian scientists pointed out that soil development, like vegetation, is governed in its gross aspect by climate. From such roots there emerged the ecosystem perspective which will be followed in this treatment of ecologic plant geography.

It has become a firmly entrenched axiom in ecologic plant geography that climate is the primary force governing the major features of the earth's vegetation, with soil and microclimate playing secondary roles, and fire or animal influence often modifying the pattern still farther. Accordingly, the major vegetation types will be characterized along with their climatic, edaphic, and successional relations, with brief commentary on the principal vertebrates involved as primary and secondary consumers, and with the principal types of land use suggested.

Owing to the enormous complexity of such subject matter when all of North America is to be considered, much thought has been directed to elaborating on a classification system that is both ecologically sound and conducive to easy mental storage and retrieval on the part of the reader.

The result has been a classification system based on the stable vegetation of zonal soils (i.e., climatic climaxes) just prior to its alteration by man. Such vegetation is used to delimit climatically complementary phytogeographic areas in which many other types of vegetation are either seral, or are stable on substrates or topography differing from those of zonal soils. This makes it possible to draw the infinite variety of climax and seral vegetation into a comprehensive framework, and to establish the closest possible relations among vegetation, climate, and soil. In addition to climatic climaxes, edaphic, topographic, fire, and zootic climaxes are recognized, these adjectives indicating the nature of unusually strong influences that determine the character of stable communities which differ materially from those of zonal soils.

This emphasis cannot be described simply as being on "potential" vegetation, for that depends on man's activities, which have resulted in grassland in eastern Nebraska and pine forest in southern Alabama, etc. It is the potential vegetation determined primarily by climate and soil in the absence of man's influence that serves as the basis of this classification.

There is no implication that climax vegetation types are or ever were the major components of vegetation mosaics, for fire, insect devastation, hurricanes, floods, vulcanism, etc., have continually interrupted successional trends, which nevertheless, began again following the interruption. As long as there is any natural or seminatural vegetation left for man to manage, emphasis on potential vegetation and its environmental controls, whether the end point of succession is desirable or not, is necessary for a fundamental understanding of the resource. In this view, the basic geographic units are categories of land defined on the basis of their biotic potentialities. The objective here has been, therefore, to interpret as well as to describe, and this has led to some arrangements different from previous classifications such as those elaborated by Shantz and Zon[359] and by Küchler[242] in connection with mapping the potential vegetation of the United States.

All the area capable of supporting the same climax plant association (with the dominants of all layers or unions closely similar throughout), whether this be climatic, edaphic, or topographic climax, is called a Habitat Type.

The entire area over which zonal soils support what may be considered the same type of (climatic) climax constitutes a Zone. Zones are areas of essentially homogeneous macroclimate as indicated by a common climatic climax, regardless of the number and extent of other Habitat Types included in its geographic boundaries. They fit together on a map as a mosaic without overlap, although the association which is climatic climax in one Zone usually occurs as an edaphic or topographic climax in contiguous Zones, where it becomes restricted to environment which compensates for the relatively unfavorable macroclimate there.

Zones in which the dominants of the climatic climaxes have had much the same geologic history, exhibit a strong thread of taxonomic continuity, and occur in climates of somewhat similar pattern are grouped into Provinces. Often it is convenient to recognize Sections of a Province.

Provinces, in turn, are grouped into Regions in which there is no necessity of any taxonomic similarity, but there is a high degree of physiognomic uniformity among the climatic climaxes, and a gross similarity of climates throughout.

To emphasize the special usages of these words commonly given other definitions elsewhere, they will always be capitalized when used as defined above.

An example of this classification perspective as it has been applied in the Washington steppe is as follows.

Steppe Region (All parts of the earth's surface which have winters with considerable freezing weather but abundant summer heat, have zonal soils that are too dry for trees, and have perennial grasses dominating the climatic climxes.)

Agropyron spicatum Province (Those portions of the Steppe Region which lie mainly in the rain shadow of the Sierra–Cascade Mountains, have rainfall mostly in winter, and have *Agropyron spicatum* and other dominants drawn mainly from the Arcto–Tertiary Geoflora.)

Northern Section (That part of the Province north of the Ochoco Mountains of Oregon, where oceanic influence is strong and large areas of steppe contain no *Artemisia tridentata.*)

Agropyron spicatum–Festuca idahoensis Zone (Those parts of the *Ag- ropyron* Province in which zonal soils had the *Agropyron spicatum–Festuca idahoensis* association as their stable vegeta- tion before the advent of agriculture.)

Agropyron spicatum–Festuca idahoensis Habitat Type (All habitats in the *Agropyron-Festuca* Province which support, or once supported, grassland dominated by these two grasses.)

Elymus cinereus–Distichlis stricta Habitat Type (Saline soils that support, or once supported, communities composed largely of *Elymus cinereus* and *Distichlis stricta.*)

Artemisia rigida–Poa sandbergii Habitat Type (Basaltic lithosols which support, or once supported, climax vegetation with these two species as the dominants.)

Stipa comata–Poa sandbergii Habitat Type (Sands that support, etc.)

Owing to limitations of space and indeed of basic information on the ecologic plant geography of much of North America, little emphasis can be given categories below the level of Province and Section, although the prac- ticality of this has been demonstrated.[104,109] Variation in the amount of information available for different areas has naturally resulted in variation in the extent of treatment here.

EVOLUTION OF NORTH AMERICAN VEGETATION[117,271]

Significance of Paleoecology for Vegetation Classification

Biogeographers have divided and redivided the land surface of North America into diverse systems of formation areas, biomes, provinces, zones, etc., and in their writings these units have usually been treated as static entities without a history. Yet it is perfectly obvious that the present landscape pattern over the earth represents no more than a cross section of a continuous change in the earth's crust, its climate, and its biota that has been operating for many millions of years. Many features of vegetation must be set against this historic background to become intelligible, since they cannot be accounted for on the basis of present environmental conditions. Paleoecology thus holds the key to many enigmas that must be resolved to make a classification of the present vegetation relatively sound.

In contrast with ecologic plant geography, the historic perspective has long dominated floristic plant geography. For example, in the mid-nineteenth century Asa Gray called attention to the strong similarities between the floras of southeastern Asia and southeastern North America, which share some 156 genera of flowering plants despite their wide geographic separation, and like others before him attributed this to an earlier intercontinental continuity of ranges that subsequently became disrupted. Toward the end of that century, C. H. Merriam attempted to classify biotic provinces in North America on the basis of whether their component species had come to occupy their present ranges as a result of migration from the north or from the south, as judged from their present centers of distribution.

The first significant attempt to make use of paleontologic data in classifying the vegetation of North America was made by F. E. Clements, who had a keen appreciation of the fossil record that had been pieced together by 1916.[75] This ecologic interpretation of fossil floras was then enthusiastically pursued by paleobotanists, especially Ralph Chaney and his students. A vast quantity of historic information has come to light since this perspective developed, so that the ancestry of modern vegetation types can be traced in considerably more detail. The picture of the past continues to unfold. Many if not most fossil beds have not yet been fully studied, misidentifications still plague even the specialists, better methods of age determination are being discovered, microfossils are beginning to get their due attention, and the data of paleoecology are becoming more quantitative.

It should be obvious that plants migrate as individual species, so that a community may be said to migrate only when most of its component species change locations somewhat in unison. This would require that the ecologic amplitudes of plant species change more slowly than does the pattern of climates on a continent, and such seems to have been the case. However, paleoecology shows that specific associations seldom retain their coherence for long. Extinctions, exchanges of species between contiguous areas, evolution, and changing combinations of environmental factors all contribute to the constant remodeling of a vegetation unit through time. For example, during Wisconsin glaciation, *Picea glauca* ranged over most of eastern United States, but usually it was not accompanied by *Pinus*, as it almost invariably is today. On the other hand, over wide areas this *Picea* grew in the vicinity of *Quercus*, whereas there is very little overlap with the range of *Quercus* today. Thus past communities may have no modern analogs, even in time as recent as the last glaciation.[438] It is, nevertheless, obvious that Cenozoic fossil floras usually do not include disharmonious mixtures of genera.[279] There is still enough coherence at the level of Regions as we see them today, that their floras can be traced far back into the past despite the alterations they have undergone.

Paleoecology is a valid perspective only if processes during past geologic time were similar to processes operating now (Charles Lyell's Principle of Uniformitarianism), and this assumption seems reasonably well grounded. It is generally assumed that a group of plants which lived together at one time in the past, and still have close relatives living together, indicate approximately equivalent environments. This assumption based on taxonomy is often buttressed by certain features of leaf morphology that are associated with modern vegetation types, and are likewise characteristic of taxonomically similar fossil floras. In other situations such as the proportions of isotopes in marine shellfish or the distribution of animals with narrow environmental requirements such as alligators and tapirs or musk ox and caribou, strengthen our interpretation of past climates based on plant groupings. Strictly geologic features can often be used to verify certain of the inferences drawn from biologic data. Gypsum and other salt deposits undoubtedly indicate aridity. Extensive beds of coarse water-worn gravel suggest nearby montane environments. Elevated and enlarged continents must have had relatively cold climates. Deposits of bauxite and hematite accumulate only in wet tropical climates. By taking all these features of the earth's surface and its biota into account, we can construct a fairly substantial picture of the ancestry and relationships among modern vegetation types.

The Cretaceous Period

Except for a few finds of uncertain status, no fossils representing angiosperms have been found in sediments older than the Cretaceous Period (Table 1). Early in that Period ferns, cycads, and conifers still dominated the earth's terrestrial vegetation. Not until detailed studies of fossil pollens were made did it become rather certain that the major development of angiosperms, and their ascendency over more primitive vasculares, really occurred in the Period. Early Cretaceous pollens are of limited variety and most closely resemble the spores of seed ferns.[119] These were followed by progressively more advanced types, in ever-increasing variety upward through the sediments that accumulated during this very long geologic period. Leaf impressions of this age have long been available, but these are less definitive than pollen, and have not been critically examined in the light of more advanced techniques based on venation patterns and epidermal characters. It is highly questionable if modern generic names of plants should be applied to even the late Cretaceous fossils, but apparently a number of modern families were represented.

Judging from taxonomic relations, the tropical* belt was quite broad in Cretaceous time, for cycads occurred in western Greenland,[232] and along even the northern coast of Alaska there was forest composed of temperate zone trees.[373]

It is generally suspected that angiosperms originated in some restricted segment of the tropical belt that surrounds the earth, spreading from there to the remainder of the belt and expanding into progressively cooler climates toward the poles. This occurred while the continents as we know them were spreading apart. Floristic plant geographers have looked closely at modern distribution patterns for possible clues to the location of that ancestral home, and this has directed attention to the mountains of southeastern Asia.[374,387] Here there are abundant representatives of both holarctic and pantropic orders of angiosperms, with a high concentration of primitive members of these groups. As an example, all seven genera of the *Fagaceae* occur there, and since the endemic genus *Trigonobalanus* combines the characteristics of both subfamilies it must be an ancient relic. This tropical montane flora intergrades with both temperate floras and lowland tropical floras by means of numerous phylogenetic connecting series. Although the lowland rain forests of that area are floristically richer than the montane flora, they do not have so many primitive forms in such a diversity of families.

Evidence from the geography of mammals, reptiles, amphibia, and fresh-water fishes seems consistent with this hypothesis that southeastern Asia was an important evolutionary center from which these groups too radiated, some reaching the Americas by way of a (hypothetical) archipelago connecting New Zealand, Antarctica, and South America, with others travelling by way of Beringia.

On the other hand geologic data cast considerable doubt on the concept that the mountains of southeastern Asia were the cradle of angiosperms. Also, some would interpret the floristic situation as indicative of the fact that continental drift brought Gondwanaland and Laurasia together in this area, creating an early mixture of primitive stocks, with the biogeographic discon-

---

*The adjective "tropical" will be used in connection with areas in which the lowlands experience rare and light frost at most. Fossils of organisms that appear closely related to modern groups confined to frost-free climates will also be referred to as "tropical" even though they may occur beyond the limits of the area between the Tropics of Cancer and of Capricorn.

"Temperate" climates are those with a regular season of frosty weather, yet having a substantial heat sum that can accumulate during either many months of warm weather (as in the Puget Sound area), or fewer months of hot weather (as in the Ohio Valley).

tinuity recognized a century ago by A. R. Wallace ("Wallace's Line") marking the actual position of the union.[331]

The basis for the superiority of angiosperms leading to their ascendency
over ferns, cycads, and conifers during Cretaceous time is puzzling. Perhaps
it is related to the fact that they have shown a higher capacity to diversify.
They have more varied reproductive apparatus, more varied perennating
systems, and wider microenvironmental specialization. A coincidence that
may be related in some way is that insects and mammals evolved as the
angiosperms spread, while dinosaurs dwindled to extinction.

So far as we can discern from fossil remains, most angiosperms were trees.
Nearly all families of vascular plants contain trees, and these are typically the
most primitive members of the family. Apparently forest, scrub, and her-

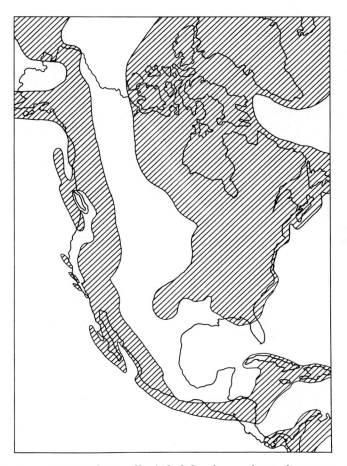

Fig. 5.   Approximate distribution of land (shaded) and sea on the North American continent
at one stage (Turonian) of the Cretaceous Period (after Schuchert[353]).

baceous types of vegetation are progressively younger, with the annual habit
latest to develop.

During the Cretaceous Period an epicontinental sea invaded North
America, extending from the Gulf of Mexico to the Mackenzie Delta, dividing the continent into two strips (Fig. 5). At the same time there appears to
have been continuous land connections between Alaska and Asia, as well as
effective continuity between northeastern North America and Europe.
Thus, from its first appearance and extending through the early part of the
Cenozoic Era, the flora of western North America was more closely similar to
that of Asia, whereas that of eastern North America had closer relations with
Europe.[232]

Eastern and western Eurasia were likewise separated by a north–south
seaway at this time.

The Paleocene Epoch

Practically throughout the known history of the earth, its surface has
alternated between two contrasted conditions. During the long Periods into
which geologic history is divided, the seas encroached upon the continents,
as in Cretaceous time, restricting the area of exposed land. In these periods
the relief was gentle, and temperature and precipitation were both rather
uniform from place to place over wide area. Such time spans were separated
by short intervals in which much land was lifted above the seas by "revo-
lutions," usually producing massive mountain ranges on the continents.
Until such rugged topography was again worn down by erosion, it resulted in
sharp differences in temperature and precipitation over land.

The climatic changes brought on by the revolution that separated the
Mesozoic and Cenozoic Periods must have elevated North America more
completely above the seas than it is today. Not only were Asian and European land connections maintained, but to the south Cuba was connected
with Central America. This uplift forced the latitudinal belts of vegetation to
retract equatorward,[372] but soon a long erosional cycle set in, and by the end
of Paleocene time climates had become warm and plant ranges had expanded
poleward again.[436] Temperate and tropical genera became intermingled
over at least most of North America, with the temperate element proportionately greater at any one latitude than it had been in the Cretaceous
Period.[54,432]

It was in this Epoch that mammals first became the dominant class of
vertebrates. The word Cenozoic means "recent animals," and the Cenozoic
Era is sometimes referred to as the "Age of Mammals."

The Eocene Epoch

During Eocene time erosion continued to wear away the land surface. A
low north–south ridge marked the area where the Rocky Mountains were
later to rise,[250] and perhaps small mountains remained elsewhere, but the

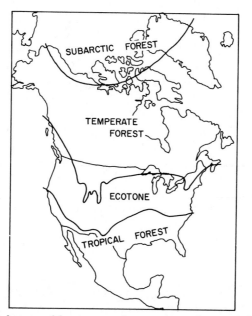

Fig. 6. General character of the vegetational pattern during the period of maximum warmth at the close of the Eocene Epoch (after Dorf[117]).

modern Black Hills, Ozarks, and Appalachians show how small mountain ranges can have little influence on the climates of the surrounding area unless they lie athwart the path of strong prevailing winds. Prevailing winds distributed moisture across the Eocene continents much more uniformly than at present.

The seas had expanded over the continental margins sufficiently that ocean currents carried warm tropical water to high latitudes, raising the temperature there. Also the increased ratio of sea to land allowed more storage of summer heat. Possibly frequent volcanic eruptions adding dust to the atmosphere also increased temperatures, or for a time, the earth may have received more heat from the sun. Although the cause is conjectural, it is certain that the climates became even a little warmer than in Cretaceous time, and by the end of the epoch even the poles must have been ice free.

Under these climatic conditions, latitudinal belts were closer to the poles, were more symmetrical across the continents, were broader, and had more gradual ecotomes than at present. A mesophytic forest similar to that which now characterizes central Canada occurred across the cold northern margin

*Plant evolution has proceeded so slowly that modern generic names are applied to Eocene fossils, and by Pliocene time fossils are treated as modern species. This stands in sharp contrast with the situation in warm-blooded animals, for modern genera in those classes do not appear until late in Miocene time, and modern species are largely Holocene in origin. Thus, animal fossils are much more critical than plant fossils in the dating of strata, but plant fossils are the more critical indicators of climate. Plants have been characterized as the "thermometers of the ages," but they are equally good as "pluviometers."

of North America, on the Arctic Archipelago and on northern Greeland (Fig. 6). Characteristic genera were *Abies,** *Betula, Corylus, Pinus, Populus, Salix,* and *Taxodium.* [32] Although the position of that forest was arctic, from its present position we are accustomed to calling it subarctic, and shall do so here. Its most significant feature is its generic impoverishment, for the same genera were also present much farther south where they played only minor roles.

Immediately south of that forest belt are found Eocene fossils of trees now characteristic of the temperate zone. This forest, like that to the north, was also mesophytic. Previously, in the Paleocene and Cretaceous forests, tropical genera outnumbered temperate nearly everywhere that fossil floras have been found, but here in the north in Eocene time the temperate group was rather well segregated. [427] The temperate element included such deciduous genera as *Acer, Aesculus, Carpinus, Carya, Castanea, Cercidiphyllum, Corylus, Fagus, Fraxinus, Ginkgo, Juglans, Taxodium, Tilia,* and *Ulmus,* mingled with evergreen conifers: *Abies, Chamaecyparis, Libocedrus, Picea, Pinus, Pseudotsuga,* and *Tsuga.* Because plants at high latitudes in the Northern Hemisphere achieved a remarkably uniform distribution around the world in Eocene time, paleobotanists have often referred to them as the Arcto–Tertiary Geoflora. Mammalian paleoecology provides rather convincing evidence that throughout the Cenozoic Era the only continuous migratory route between North America and Eurasia has been Beringia. But the continuity of early Cenozoic plant fossils across the North Atlantic is equally strong evidence that the straits separating Greenland, Iceland, Spitzbergen, the Faraoes, and northern Europe were quite narrow. Possibly ocean-floor spreading has subsequently widened these interisland gaps.

Owing to weak climatic differentiation, the temperate forest flora on the north gradually gave way to a mainly tropical flora which prevailed along the southern margin of North America. In the Wilcox flora of the latter region such temperate genera as *Alnus, Betula, Castanea, Corylus, Nyssa, Pinus,* and *Sassafras* grew in close association with tropical genera such as *Cycas, Drimys, Engelhardtia, Ficus, Laurus, Nectandra, Persea, Sabal, Sterculia, Tetracera,* and *Xylosma.* Eighty-three percent of that flora had entire leaves—a proportion which closely approaches the character of the Amazon Valley today (88%)—and lianas were well represented. Crocodiles extended as far north as New Jersey, alligators to Wyoming, and coral reefs formed as much as 2400 km farther north than at present. Along the Pacific coast an ever-narrowing strip of this mixed tropical and temperate flora with palms, *Artocarpus,* and cycads extended to 62° north latitude. [434]

During this epoch the more purely tropical forest (composed of the Neo-Tropical Geoflora) became disrupted, for Cuba lost its connection with Central America and the Isthmus of Panama sank to form a salt-water gap separating the biotas of North and South America that was to persist until Pliocene time.

The pattern of the circulation of the earth's atmosphere includes a perma-

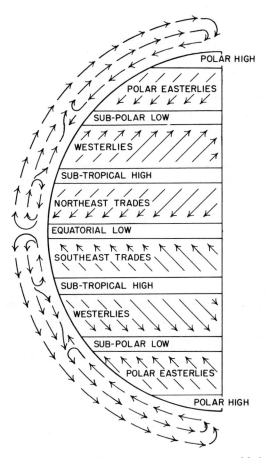

POLAR HIGH

POLAR EASTERLIES

SUB-POLAR LOW

WESTERLIES

SUB-TROPICAL HIGH

NORTHEAST TRADES

EQUATORIAL LOW

SOUTHEAST TRADES

SUB-TROPICAL HIGH

WESTERLIES

SUB-POLAR LOW

POLAR EASTERLIES

POLAR HIGH

Fig. 7. Schematic representation of major air mass movements and belts of high and low pressure over (half of) the earth.

nent high pressure belt centered approximately on the Tropics of Cancer and of Capricorn (Fig. 7). At these latitudes the climates are dry as a result of low rainfall combined with strong insolation (Table 2). In this belt dryness is most intense at the western margins of the continents, resulting in deserts at the present time, with aridity diminishing eastward until tropical xerophytic forest prevails at the eastern margins of continents in the northern hemisphere. If in Eocene time continents were in the same latitudes as they are today, some degree of aridity was undoubtedly present in these belts. But aridity does not favor fossilization, so direct evidence of strongly xerophytic vegetation has not been found there in Eocene time.

There is no sound basis for assuming that the evolution of xerophytes has been much more rapid than for other groups, so the highly specialized xerophytes now distributed in widely separated centers of aridity strongly suggest the antiquity of the dry environments that they needed to evolve.

TABLE 2

Mean Annual Precipitation/Evaporation Ratio by Latitude,
as Calculated from Data Presented by Sellers[357a]

Latitude	P/E ratio
80–90°N	2.86
70–80°N	1.28
60–70°N	1.25
50–60°N	1.68
40–50°N	1.41
30–40°N	0.87
20–30°N	0.63
10–20°N	0.83
0–10°N	1.57
0–10°S	1.11
10–20°S	0.73
20–30°S	0.61
30–40°S	0.74
40–50°S	1.37
50–60°S	2.01
60–70°S	2.40
70–80°S	1.82

[a] Although differences in the relative amounts of land surface, elevation, and cloudiness distort symmetry, the subtropical dry belt centered approximately on the tropics at 23.5° latitude is clearly marked by evaporation exceeding precipitation.

For example, *Mendora*, a desertic genus, occurs near the Mexico–United States border, then reappears in the dry subtropical belts of South America and Africa, with *Hoffmanseggia* showing essentially the same pattern. These and other disjunctions among xerophytes have extremely low probability of having developed in short time. They may even date back to the Cretaceous Period when continents had not yet completely lost connection.

An Eocene flora from northeastern Nevada, consisting entirely of temperate zone plants, appears to represent relatively xerophytic woodland on a low ridge in the area where the Rockies were later to rise.[19] A little farther east in southern Wyoming is a fossil flora suggesting savanna.[250] Then too, the development of ungulates starting in this Epoch may suggest more than just local parks in the generally forested landscapes.[399] All these fossils may collectively represent the fringe of an arid area, the center of which was to the southwest near the Mexico–United States border where aridity was too severe for fossilization. The xerophytic flora that probably occurred in that area, or was to develop there later, has been referred to as the Madro-Tertiary Geoflora, since the Sierra Madre Occidentale in northwestern

Mexico is somewhat centrally located with respect to the astronomically determined center of aridity.

The Oligocene Epoch

In Oligocene time a drop in temperature began that was to continue through subsequent epochs and culminate in Pleistocene glaciation (Table 1). Although temperatures fluctuated during that long time span, the overall trend was that of progressive cooling, with the broad transcontinental belts of vegetation moving equatorward along with the isotherms controlling them.

Cooling in Oligocene time was especially rapid.[254,436] It is suspected that this involved winter temperatures more strongly, since tropical elements made the more spectacular retreat, withdrawing to the southern margin of the United States. The strictly temperate residue left in the north still had full access to land connections with Asia across Beringia.[159,435]

The Rocky Mountains had uplifted somewhat in the geologic revolution that terminated the Mesozoic Era, but there is no evidence that this was sufficient to result in a pronounced rain shadow. However, in the Oligocene Epoch the low ridge that remained rose to sufficient height that rainfall declined markedly to its leeward (eastward). Forest thinned on the interfluves in this rain shadow, leaving preadapted xerophytes to form savanna.[117,271] New combinations of climatic and edaphic conditions, the merging of islands of xerophytism formerly isolated on dry microsites, and reduced competition probably operated synergistically to speed up the evolution of xerophytes at this time. Herbs, previously rare in the fossil record, began to gain rapidly in representation, and the herbaceous life form was a necessary precursor to the development of annuals which were to become so characteristic of dry climates. Despite the extremely unfavorable conditions for fossilization in arid regions, some fruits of the ancestors of *Panicum*, *Setaria*, and *Stipa* have been identified in Oligocene strata east of the Rocky Mountains.[125,270] These grasses had developed short basal internodes or rhizomes, basal leaf meristems, and buds protected either by their position beneath the ground or by persistent leaf bases—characters that enabled them to take advantage of the abundant solar energy available in environments too dry for a continuous tree canopy, and at the same time minimize damage from grazers. Grasses of this type, together with fossils representing a few other xerophytic genera, by themselves provide substantial evidence of the appearance of vegetation at least as open as savanna. But independent corroborative evidence is provided by the synchronous rapid rise in the fossil record of cursorial horses and camels with long necks and jaws that coincided with a decrease in browsers such as the tapir.[125] Although the horses, camels, and rhinoceroses of that rain shadow developed the high-crowned teeth characteristic of grazing animals that depend on grass leaves with their harsh, silicified epidermis, browsers persisted too, suggesting that broad strips of gallery forest persisted along rivers.

Uplands in the Rockies had temperate forests, with temperate and tropical genera growing intermingled on the basal plains outside the rain shadow area.[427]

The Miocene Epoch

Early in the Miocene Epoch temperate mesophytic forests composed of mixtures of conifer and angiosperm trees (*Fagus, Liquidambar, Quercus, Tilia, Abies, Chamaecyparis, Picea, Thuja, Tsuga,* etc.) extended from Alaska to Oregon on the lowlands, with the coniferous elements prevailing on the highlands. But by the end of that epoch the angiosperm element had receded from Alaska leaving conifers in essentially complete control there.[435] Beringia developed a water gap during this epoch, so the subarctic forest which had not yet claimed that area, was to arrive too late to make floristic exchanges with the homologous subarctic flora of Asia.

As uplift continued along what is now the main axis of the Rockies, aridity intensified in the rain shadow immediately to their leeward. Savanna was displaced eastward,[117] out of the most severe part of the rain shadow next the mountains, and was replaced there by steppe. The commonness of fossils of both *Stipae* and *Panicae* in Miocene sediments east of the Rockies attests to the abundance of xerophytic grasses, either as dominants of steppe, or as the matrix of savanna.[125]

By this time the extent and continuity of the midcontinental arid region had become sufficient to effectively split the tide of mesophytic forest species which had been moving southward in response to continental cooling. A broad lobe continued southward to the east of the arid area, and a western segment threaded its way southward along the Pacific coast and along the moist slopes of the Rockies and low hills in the Cascade–Sierra area. Modern birds and amphibia as well as plants reflect this decisive splitting of temperate mesophytic biotas into eastern and western segments in mid-Cenozoic time.

The low hills that until now had marked the Sierra axis began to uplift in Miocene time, and the rain shadow developing to their leeward gave impetus to the northward expansion of more extreme xerophytes from the old center of the Madro–Tertiary Geoflora. Chaparral spread northward across what is now the Great Basin area. The Mojave basin had not yet become desert, but it had developed a microphyllous woodland. This expansion of aridity west of the Rockies greatly complicated the southward migration of the western arm of the temperate mesophytic forest.

As drouth was intensifying in places, summers were becoming increasingly more cool. All these climatic stresses probably stimulated further evolution from the arborescent to the frutescent to the herbaceous life form. Herbs have several advantages under such stresses. They can devote a greater proportion of their production to seed formation, and they can migrate faster because shorter time is required to start flowering. What is most important of all, in dry climates herbs can aestivate or survive only as seeds in seasons when soil moisture becomes unavailable—a capacity which nonsucculent woody plants lack. Prior to Miocene time there is limited evidence for the existence of herbs, but now they began appearing in abundance, and continued to increase in diversity through the succeeding epochs.

The Eurasian continent had no north–south mountain axis to create a major rain shadow, but that land mass is very large and the central area has low rainfall simply because it is so far from oceanic sources of moisture. There too an intensification of mid-continental dryness apparently caused the southwardly retreating temperate mesophytic flora to split into western and eastern segments, which, thereafter, remained completely isolated from each other.[147]

The Pliocene Epoch

Cooling continued through the Pliocene Epoch, and it has been estimated that by the end of the epoch mean annual temperatures had declined about 15°C below the warm-maximum of Eocene time.[117] Vegetation belts became better differentiated, and moved about as far southward as they are today, with tropical plants becoming restricted to the southern margin of North America. From the general pattern of the vegetation it might be inferred that glaciers had already developed at high latitudes and altitudes, and that there existed sizeable areas too cold for trees (i.e., tundra) at both high latitudes and high altitudes.

By mid-Pliocene time the Rockies lacked only 300–600 m of their present height,[341] so the rain shadow to their leeward was nearly as pronounced as it is at present.[271]

The general elevational increase of the land resulted in the elimination of the long-standing salt water barrier to intercontinental biotic exchange along the Isthmus of Panama. But at the northern margin of the continent the reverse happened, for it was not until Pliocene time that arctic land connections with Asia were severed. It was during this epoch that forest gave way to tundra in Alaska.[435]

West of the Rockies another important event took place. The Sierra–Cascade axis, which up to now had consisted mainly of low hills capped with temperate coniferous forest, began to gain height progressively by accumulating volcanic debris or by faulting. This interposed a new barrier across the path of air masses moving from west to east across North America, and moisture interception by these mountains dried up the trough between them and the Rockies. Most species of the chaparral that had spread earlier

up through the Great Basin area now retreated southward and were replaced
by xerophytic steppe plants more tolerant of the colder, drier climate.

The principal plants of this new area of steppe included species in *Agropyron*, *Artemisia*, *Atriplex*, *Festuca*, and *Poa* that must have been occupying
special habitats in the temperate forest, for their importance and close similarity in North America and Eurasia suggest high latitude ranges during early
Cenozoic time which gave them access to the Bering bridge.[17] Other
xerophytes in the newly arid intermountain trough, such as *Chrysothamnus*,
Gutierrezia, *Hilaria*, *Purshia*, and *Tetradymia*, must have evolved from tropical ancestors and as such have affinities to the south. They are not represented in Eurasia presumably because Beringia was beyond their climatic
tolerance even at the height of Eocene warmth. It was during this Pliocene
Epoch that a number of herbaceous plants make their first appearance in
western sediments as pollen grains.

With the lowlands becoming so arid, and the mountains remaining moist
but intolerably deficient in summer heat, temperate forest elements in the
west lacked the opportunity to move freely southward like their homologs
which had become segregated on the east side of the midcontinental steppe.
As a consequence, temperate mesophytic forest in the west lost most of its
angiosperm dominants and became extremely impoverished of tree species
except along the cool seaward slope of the new mountain axis.[253] Angiosperm trees such as *Carya*, *Castanea*, *Fagus*, *Liquidambar*, *Liriodendron*,
Tilia, and *Ulmus* that had still been well represented in west central North
America in Miocene time, now became extinct in the west, leaving their
coniferous associates in control: *Pseudotsuga*, *Tsuga*, *Chamaecyparis*, *Thuja*,
Sequoia, etc. Some have hypothesized that a decline in summer rainfall was
responsible for this loss of broad-leaved trees, but the coast of northern
California remained much wetter than areas along the one-hundredth merid-
ian, yet conifers replaced angiosperms in the former area, with angiosperms gaining control in the latter.

About the southern Rockies, in the part of the Subtropical High Pressure
belt which had long been the cradle of the Madro–Tertiary Geoflora, climates had also been becoming drier. Successively more xerophytic types of
vegetation had been evolving there and spreading outward as centrifugal
waves, to be replaced each time by still more drouth-adapted types in the
core of the area. Apparently, it was not until Pliocene time that this core area
became dry enough that even steppe receded, allowing those residual plants
that had long been restricted to the most severe habitats of the area to come
together as a desert vegetation.

With nearly one-fourth of the continent now arid or semiarid, xerophytes
had excellent opportunities to achieve wide ranges, just as temperate zone
mesophytes had had their greatest opportunity earlier in Cenozoic time.
Annuals probably evolved rapidly as deserts came into existence, for their
ability to survive entirely by seeds confers superb capacity to endure long

and variable periods of drought with minimal hazard. Then, it enables the individual to develop in direct proportion to the amount of rain when it does come, thus exploiting available moisture to the maximum as it produces the new seed crop.

These drastic consequences of aridity resulting from north–south mountains in western North America had no counterpart in Eurasia. The east–west chain of mountains extending from the Pyrenees to the Himalayas was too far south to interfere greatly with the Pliocene migration of temperate mesophytic forest, and east of the Himalayas there was even less obstruction to compensatory southward migration.

The impact of the long trend of cooling during the Cenozoic Era on migration and evolution is reflected clearly in the changing ratios of woody to herbaceous species in what are now temperate latitudes. In England 97% of the Eocene plant fossils represent woody plants. By Pliocene time this had dropped to 22%, with subsequent extinctions during the Ice Ages reducing it to 17%.

The Pleistocene Epoch[112,147]

By the end of the Pliocene Epoch the average elevation of the land had risen about 500 m above the low point of late Eocene time. This could have made the climates very responsive to cyclic variations in the amount of heat received from the sun, resulting in the series of glaciations which then occurred.

Each glacial episode in the Pleistocene Epoch involved a period of slowly dropping temperatures with ice accumulating and spreading, followed by a rather rapid rise in temperature that melted the ice and culminated in a warm interglacial period. The progressive accumulation of snow during glacials is suspected as having been simply a consequence of summer temperatures dropping below a critical level so that melting did not equal snowfall. Once a snow cover started to persist through summer, its reflectivity would return even more solar radiation back into space, consequently feedback would favor more snow accumulation. As it accumulated, the snow compacted into ice, and when the mass became thick enough toward the center, pressure caused it to begin flowing imperceptibly outward in all directions.

The cooling was global, as shown by evidence of mountain glaciation of Pleistocene age even in the tropics. In the Caribbean area temperatures dropped 7–8°C below interglacial levels,[128] but this was not enough to reach the critical frost point, so tropical vegetation persisted there throughout.

In North America each major ice field began accumulating well within the land area of Canada, spreading centrifugally until the cooling trend reversed. As glaciers expanded northward to the Arctic Ocean the tundra belt was obliterated except on a few projecting headlands along the coast that were not covered by ice. Spread to the south was less disastrous to plants and animals owing to the excellent opportunities for southward range displacement.

There appear to have been at least seven glacial periods in the Pleistocene Epoch,[128,356] but the scouring action of one glacier virtually destroys all evidences of previous glaciation in the area it covers, and only the last few can be recognized in terrestrial features. On the North American continent only four major periods of ice accumulation and spread are generally recognized, these evident around the margin of the glaciated area where subsequent scouring did not erase all of the edges of previous glacial deposits. The last glaciation, the Wisconsin, began about 170,000 BP (Before Present), and at its zenith approximately 30% of the earth's surface was covered with ice. Although most evidence of biogeographic events associated with pre-Wisconsin glaciation was eliminated, it is probably safe to assume that earlier glaciations had much the same influence on life as the last, since all four extended about the same distance southward over the continent. Each glaciation was accompanied by a temporary shift of species to lower latitudes, or to lower altitudes in the mountains. Then as the ice melted away, the vacated land was recolonized by range expansions from the south or from lower altitudes, and by radial migration from locally unglaciated refugia that had been surrounded by ice. The interglacials were long enough and warm enough for biotas to move poleward until they reached latitudes even higher than at present. The last interglacial must have been 2–3°C warmer, for the tapir migrated as far north as Indiana and Ohio, and the manatee inhabited the Atlantic coast as far north as New Jersey.[184]

During glaciation so much water became tied up as ice on land that sea levels dropped about 140 m, exposing broad strips of thinly vegetated coastlines along which organisms could migrate with ease. Land became continuous across the Bering Strait since the water now has a maximum depth of only 55 m, and it was during Wisconsin glaciation that despite the formidable (tundra) climate, man crossed this ice-free area to become established in the Western Hemisphere.

During interglacials when temperatures were higher than at present, sea levels rose up to 30 m higher than they are now. Strips of marine sands along present shorelines were deposited during those submergences, to become the infertile parent materials for soils that now support highly distinctive edaphic climaxes.

Although each glaciation covered about the same area as the preceding one, the degree of continentality of climates is believed to have increased progressively. Many large mammals that became restricted if not extinct at the close of Wisconsin glaciation may have suffered from the effects of rapid warming from the very cold conditions of late glacial time, but some believe that overkill by early man caused their demise.

Beyond ice limits the lowered temperatures reduced evapotranspiration so that the same amount of rainfall became more effective. Frequently the word "pluvial" has been used for the moister climates beyond the glaciers, but this is misleading for there is no evidence that rainfall increased. Owing to the melting of accumulated ice and reduced evapotranspiration, two vast lakes accumulated in the Great Basin—Lake Lahontan and Lake Bonneville. The former evaporated away subsequently, with the present Great Salt Lake in Utah representing a remnant of Lake Bonneville.

Glaciations appear to have been concomitant on all the continents. In Europe a chain of glacier-clad mountains consisting of the Pyrenees, Alps, and Caucasus, held in the cold air flowing from the north and prevented its dilution with warm air from the south. Ice from the north approached to within 480 km of mountain glaciers that expanded from the Alps and other mountains of southern Europe. Thus, a pocket of frigid climate suitable for hardly more than tundra species developed to the north of those mountains. Many temperate zone trees (e.g., *Carya, Liquidambar, Robinia, Tsuga*) became extinct in Europe either in Pliocene or Pleistocene time, with survival south of the mountains along the Mediterranean coast limited by dryness. In the middle of the Eurasian continent aridity continued to exclude forest, and in that area tundra telescoped with steppe during glaciations.

In eastern North America there were no transverse mountain chains to prevent warm tropical air from moving northward and diluting cold air flowing off the ice, or to impede the southward dispersion of plants and animals. In consequence, the climate remained relatively mild over a large area south of the Missouri and Ohio Rivers, and a relatively rich flora survived the late Cenozoic glaciations. Close to the ice coniferous trees dominated, but hardy species of the temperate forest were not eliminated. Southward the proportion of conifers decreased as broad-leaved trees increased.

The cooler temperatures of glacial climate reduced aridity in the midcontinental area so that forest encroached upon steppe extensively, although perhaps occupying only the moister sites in a forest–steppe mosaic. *Picea glauca* extended southward almost to the Gulf of Mexico, and at the same time deciduous forest species spread westward to the foothills of the Rockies.

The Holocene Epoch

The Wisconsin glacier reached its southernmost extension about 16,000–18,000 BP, and this was followed by a long period of minor retreats and advances. About 10,500 years ago the climate began to warm rather abruptly, and the ice margin began a rather continuous retreat—the start of the Holocene Epoch. In North America this allowed forest to advance over the relinquished territory at about 30 km a century, the same rate as estimated for Europe.[138,296] This rate of advance during deglaciation suggests that climatic amelioration was so rapid that only the innate capacity of the trees to migrate was limiting. Since this northward migration involved establishment on bare or at least thinly occupied land, it was much more rapid than the southward retreat ahead of the expanding ice had been, for that had involved infiltration into closed communities and meeting competition from established plants. Throughout the northern hemisphere it was various species of *Pinus* which led the northward advance of forest.

Aside from progressive deglaciation, the most outstanding environmental phenomenon of the Holocene Epoch was that the major trend of warming which started the wastage of Wisconsin ice continued for several thousands of years until temperatures were 2–3°C warmer than at present, and melting ice raised sea levels above those of the present time. Then temperatures declined irregularly toward present levels.

The span of time when temperatures were significantly warmer than at present has been variously referred to as the Xerothermic, Altithermal, Hypsithermal, or Atlantic Interval. Among these the first is most meaningful from the standpoint of ecology, since precipitation did not increase in proportion to heat, but the third is most popular usage at present. In consequence of decreased precipitation effectivity, xerophytic vegetation expanded in temperate and tropical latitudes, stream discharge decreased, and basins now naturally flooded accumulated a layer of loess during this time. The Hypsithermal Interval has no well-defined beginning or end, and the estimated date of its zenith is apparently not the same everywhere; radiocarbon dates indicating approximately 7000 years BP in west central North America, with later dates toward the eastern margin of the continent.

While temperatures were elevated both arctic and alpine timberlines rose temporarily to latitudes or altitudes previously too heat deficient to support arborescent plants.

During the general cooling trend of post-Hypsithermal time there have been many small reversals. During three abnormally cool periods mountain glaciers advanced, but during the last century global temperatures have risen, especially the winter temperatures of high latitudes. From 1910 to 1950 mean winter temperatures in Spitzbergen (10–12°N lat.) rose 8°C, although the only change at the equator seems to have been a slight rise in the elevation of snowlines in the mountains.[429] Between 1930–1948 the melting of glacial ice raised sea levels by as much as 15 cm, and there is still enough ice on land that if all of it were to melt, sea levels would rise sufficiently to flood all the world's major cities! At the arctic timberline forest has again been advancing vigorously onto tundra, and in temperate latitudes both marine and land species have been expanding poleward. Thus, climatic fluctuations of varying direction and magnitude are superimposed, and future trends are wholly unpredictable.

According to the timing of alternations of Pleistocene glacial and interglacial periods, the Hypsithermal Interval may well have been the peak temperature of another interglacial, with the long-term trend being toward another glaciation despite the recent warming. Interglacials have been estimated to have been 10,000–30,000 years long, so with post-Wisconsin time now 10,500 years along, and with temperatures having risen to a maximum 7000 years ago, viewing the Holocene Epoch as another interglacial seems a likely outlook.

SOUTH AMERICAN VEGETATIONAL HISTORY[228,284]

Paleontology indicates that the Cenozoic history of South America paralleled that of North America. Early in the era a temperate mesophytic *Nothofagus–Araucaria* forest (Antarcto–Tertiary Geoflora) occupied the southern tip of South America, which does not extend far enough south to have climates as cold as subarctic North America. This temperate forest was continuous, or had been continuous, by way of island links across the South Pacific, with similar vegetation in the New Zealand area. Some additional representative genera in that forest were *Coprosma, Dacrydium, Drimys, Fitzroya, Pittosporum, Podocarpus,* and *Saxagothea.*

To the north of the temperate forest was a tropical mesophytic *Maytenus–Zamia* belt. During Cenozoic time progressive cooling resulted in an equatorward shift of the major ecotone between these two vegetation belts, with concomitant climatic diversification allowing the development or at least the expansion of a large arid region, so that both types of mesophytic forest were reduced in area.

Although the temperate forest belts of both North and South America shifted toward the equator during the Cenozoic Era, they have always been kept separate by the tropical forest and so have not been able to pool their floras. This is not equally true of the xerophytic vegetation types, however. The uniformly high temperatures of the equatorial belt were not limiting for such genera as *Bouteloua* and *Larrea,* for they appear to have used locally dry habitats as stepping stones, gaining representation in arid regions on both sides of the tropical belt. Only a few temperate zone mesophytes were able to cross the tropics along the mountainous cordillera, possibly because most of them require climates with alternating cool and warm seasons, and the tropics do not provide such at any latitude.

Pleistocene dislocations of vegetation types in South America were a mirror image of those in North America, and as in North America there are disjunct communities poleward from their major ranges which indicate temporary Hypsithermal range expansions.

The foregoing provides a brief outline of the major trends in the development of Western Hemisphere vegetation during the period of angiosperm dominance. Further details will be elaborated as each unit of modern vegetation is discussed.

THE TUNDRA REGION[35]

The Tundra Region is coincident with vegetation dominated by perennial herbs, shrubs, and/or cryptograms, occurring at high altitudes or latitudes where the heat supply is too low to allow the development of the tree life form.

Tundra is a Russian word originally referring to only the marshy plains of northern Eurasia, but it has been assimilated into the English language and its use has gradually been extended to include low latitude montane vegetation above tree limits too.

Inadequacy of heat for plants with the tree life form is a common denominator of tundra environment throughout the world. Apparently as the supply of heat energy declines, the plant can afford fewer and fewer of the nonphotosynthetic living cells which make up so much of a tree. Stem tissues become especially reduced, so that most of the shoot consists of leaves, and these leaves are consequently located in the layer of relatively warm air next the ground.

In the Northern Hemisphere the average position of the Arctic Front in summer appears to determine the southerly boundary of the transcontinental tundra belt. This has long been quantified in an ecologically meaningful manner as the isotherm representing a mean temperature of the warmest month at approximately 10°C. Since on mountains at lower latitudes the position of this isotherm is obviously not determined by the Arctic Front, the fact that approximately the same temperature relations hold even in the tropics strongly suggest a fundamental relationship between the tree life form and the quantity of heat necessary to support it.[100]

In addition to being of low stature, tundra vegetation is everywhere xeromorphic. Annuals are usually rare. The cushion and rosette life forms are very common among the perennials.

The major subdivisions of the Tundra Region that will be recognized are (1) Arctic and Antarctic tundra, beyond the latitudinal limits of tree growth, and (2) extratropical and (3) tropical tundras which are above the altitudinal limits of tree growth at lower latitudes. Treatment of the last will be deferred to that section of the book that considers all tropical vegetation.

Northern Extratropical Tundras

Origin and Distribution[220,260]

The geologic record reveals no evidence of the existence of tundra climate or vegetation during the warmer epochs of Cretaceous and early Cenozoic time. However, preadapted ancestors of modern tundra plants may have had a long history in polar regions in special habitats provided by frost pockets, snow-patches, bogs, and ravines well shielded from the sun. Other tundra plants such as *Empetrum, Ledum,* and *Rubus* may represent preadapted forest species that persisted when climates cooled and trees receded from high latitudes and altitudes. This has rather certainly been a major source where tundras developed independently on isolated mountains at low latitudes, and some of these alpine species probably had opportunity to migrate northward to become incorporated into arctic tundra too. Pronounced physiologic and morphologic adaptations of modern tundra plants, the numerous widespread endemics, and the high proportion of polyploids all suggest that these plants are not particularly recent in origin.

In North America it is likely that a coherent tundra belt developed first as a fringe across the northern border of the continent in mid-Cenozoic time, then spread southward as cooling continued, eventually meeting and intermingling with local tundra floras that had developed independently on mountain summits. Since subarctic forest had extended as far south as Lake Baikal in Europe by Miocene time, the northern border of Eurasia probably supported a belt of tundra by then.[169]

The dropping of sea levels as Pleistocene ice accumulated exposed much of the continental shelves and decreased the water gaps between continents and islands. At this time, if not before, there was achieved a striking floristic continuity even at the species level among the tundra floras of North America, Eurasia, and the intervening islands which are now widely spaced. But soon the expanding ice sheets hampered all but local migrations, and eventually they covered nearly all the area that had been occupied by tundra.

Since the centers of Pleistocene ice accumulation in North America were south of the transcontinental tundra belt, this drastically limited access to the narrow belt of periglacial climate that developed along the southern edge of the ice caps. As ice expanded in all directions from centers of accumulation, plant life to the north was trapped between the Arctic Ocean and advancing ice. In this way the ranges of tundra species were drastically reduced in the north, with species persistence limited to ice-free refugia scattered along the coast of the Arctic Ocean, to headlands projecting into the Atlantic and Pacific Oceans, to a large area in the Yukon Valley, and to nunataks. Modern disjunctions and endemism clearly reflect all these types of refugia.

In the mountains at low latitude these frigid episodes allowed the alpine belts to extend downslope 100 m or more, which permitted alpine species to cross lowlands separating the mountain summits on which they had previously been isolated. To the resultant amalgam were added those arctic species that were able to come southward ahead of the ice. The ubiquity of opportunities for floristic homogenization is reflected in the ranges of *Oxyria digyna*, *Silene acaulis*, *Trisetum spicatum*, etc., which are circumpolar and reappear in almost all the major mountain ranges of temperate latitudes in the Northern Hemisphere. Some of these widespread species may have had their ultimate origins on mountain summits at low latitude and migrated northward; others spread by the reverse route.

In central Eurasia tundra telescoped with steppe as the continent cooled, but in central North America ice appears to have advanced directly onto steppe after overriding the subarctic forest belt.

In Europe periglacial climates were severe. The bottom layers of sediments that accumulated in ponds formed as the last continental ice receded contain no tree pollens. Well up from the bottom fossils of cold-tolerant conifers become abundant, with these eventually giving way, in turn, to pollen representing the dominants of temperate mesophytic forest. From such stratigraphic evidence we can infer that a broad belt of tundra occurred in front of the ice, and followed it as the ice receded.

Despite the abundant evidence of Pleistocene frost-churning of soil up to 160 km beyond the glacial border in North America (Fig. 8), pollen assemblages interpreted as indicating essentially treeless vegetation always contain some tree pollen, so either there were individual trees scattered over the landscape immediately in front of the ice, or forest was not far away.[241] Also, in southwestern Wisconsin temperate forest herbs persisted in habitats of compensation while ice lobes flowed past their relatively small unglaciated refugium, indicating that periglacial climates were not extremely severe.

Most of the tundra flora that managed to migrate southward ahead of the Wisconsin ice sheet in eastern North America appears to have existed in a belt just in front of the ice on the Atlantic coast, with a peninsular extension

Fig. 8.   "Fossil" stone net near Davenport, Washington, approximately 2000 m below the altitude of alpine vegetation at present. This site was at the border of Wisconsin glaciation.

down the Appalachians. During deglaciation tundra plants may have tended to spread westward on the newly exposed till, but north of the Great Lakes the first vegetation to invade this till appears to have been less of a tundra than of a wet meadow with cold-tolerant grasses in *Agropyron, Elymus, Festuca,* and *Poa.* In places there is indication that a belt of shrubby *Betula* followed the retreating margin of tundra as it shifted northward.[438]

Tundra animals as well as tundra plants wandered far south. Lemmings reached Mississippi, Texas, and southern California; walrus was at home on the Georgia coast.[183,184]

Although Wisconsin glaciation undoubtedly had the effect of increasing the homogeneity of the arctic–alpine flora by lowering timberlines and bringing tundra areas closer together, the Hypsithermal Interval had the opposite effect since only the highest mountains remained climatically suitable during that warm period. In the arctic a similar purge took place at this time, for arctic timberlines advanced 280–350 km north of their present positions, crowding plants against the Arctic Ocean.[55,344]

Isolation while ranges were so fragmented, combined with the stresses of Pleistocene climate changing in alternating directions, undoubtedly stimulated evolutionary changes among those tundra species that were not eliminated.

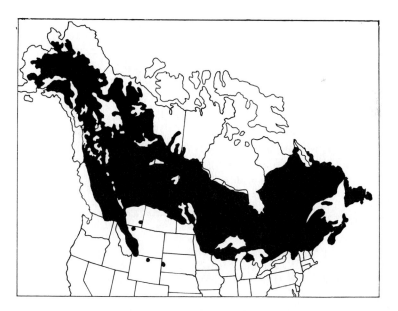

Fig. 9. Range of *Picea glauca,* including hybrids with *P. rubens* at the southeastern extremity, and with *P. engelmannii* at the southwestern extremity. Since this is the most ubiquitous tree in the subarctic forest of North America, the northerly limits of its range coincide rather well with the southerly limits of arctic tundra.

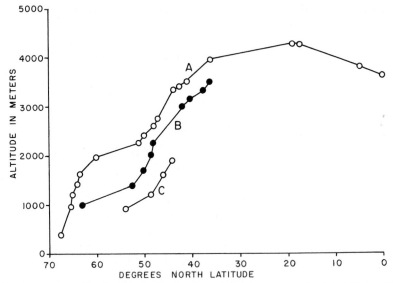

Fig. 10. Upper altitudinal limits of trees along mountain axes, as related to latitude. These are the highest reported occurrences of species that have the tree life form at lower altitudes. A, Cordillera well inland from the Pacific Ocean; B, mountains near the Pacific Ocean from southern Alaska to southern California; C, Torngat Mountains to Labrador to northeastern United States.

The largest unit of tundra at present is a broad circumpolar belt in the Northern Hemisphere, the arctic tundra (Fig. 9). In the Southern Hemisphere there is very little land area in the corresponding climatic belt, consequently only a small part of Antarctica and a few isolated islands lie beyond the climatic limits of tree growth.

The term alpine tundra is used here for all vegetation of high mountains between arctic and antarctic timberlines, wherever such mountains rise high enough that inadequate heat excludes plants with the tree life form. In the Americas the lower ecotone of alpine tundra lies at approximately 3600 m at the equator, rises to 4200 m at 20°N latitude, then descends more or less gradually to the arctic timberline (Fig. 10). Where tree lines can be com-

Fig. 11.   Seasonal trends of mean monthly temperature and precipitation at stations representing vegetation Provinces or Regions are presented. (See also Figs. 12–14.) Median precipitation values would have been much more meaningful ecologically, had they been available. Temperature is shown with a solid line, precipitation with a broken line. Since monthly precipitation in excess of about 140 mm represents useless excess in any climate, whereas differences in mean temperature below 0°C are of little significance in plant geography, the graphs have been truncated accordingly. The scales are chosen only with a view to bring the two lines close together, and the reader is cautioned against assuming the drought is indicated by the precipitation line dropping below the temperature line. The significance of a short drop below is variable depending on the amount of previous precipitation as this determined the amount of water stored in the soil, and depending on the absolute level of temperature as this indicates the amount of heat available to vaporize water.

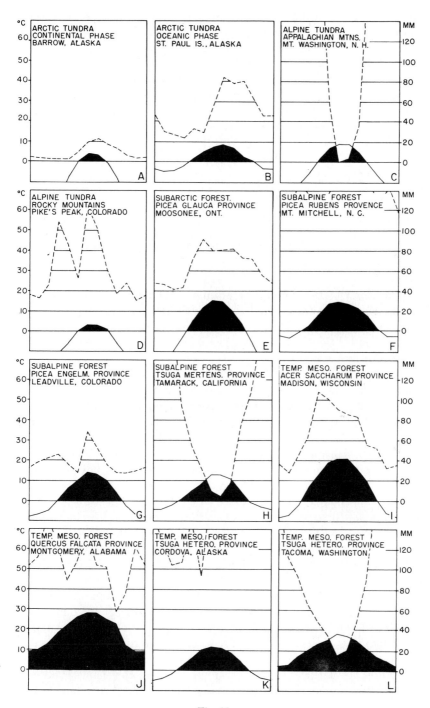

Fig. 11.

pared at the same latitude, they are lowest in the Appalachians, higher in the Cascades, and highest in the Rockies. The altitude of timberline rises higher the greater the mass of the mountain, or the higher the altitude of the basal plain. With temperature as the major determining factor, aspect is important too, as this influences heat supply, with trees ascending higher on equatorward slopes than on poleward slopes.

At the present time alpine areas are for the most part highly disjunct, the more so the farther from the poles. In consequence, endemism increases with distance from the poles.

The Arctic Tundra[45,227,321,390]

Environment. Above the Arctic Circle midwinter is sunless and midsummer nightless. Consequently, even in continental climates there is relatively little spread between day and night temperatures either in midwinter or midsummer. Despite the low angles of the sun, long daylengths in the arctic summer provide greater quantities of radiant energy at the earth's surface in each 24-hour period than reach the surface in the tropics under 12-hour days.[229] Also while the sun is above the horizon 24 hours a day the soil surface remains warmer than the air so that although mean daily temperatures as measured in standard instrument shelters continue to decline poleward across the arctic tundra, the temperatures near the ground do not decline correspondingly.[79]

Mainly as a result of the inability of cold air to hold much moisture, precipitation at high latitudes is far below the average for the earth's surface as a whole (Fig. 11A,B). In the coldest part it is indeed so dry that vegetation occurs mainly in the drainageways from snowbanks.[351] Over most of the area, however, low temperatures make the limited supply of water so effective that the precipitation/evaporation ratio is very high (Table 2), so the substrate remains moist throughout summer, with the air having a relative humidity usually in excess of 70%. In fact, over vast areas of low relief there is a period of about a month in spring when the slow thawing of the surface soil, coupled with the impermeability of the subsoil, turns the landscape into a marsh—the "marshy plain" indicated by the word tundra. Where tundra occurs on islands, or on coastlines with prevailing on-shore winds, the climates are more oceanic and have higher rainfall in comparison with inland locations (Fig. 11A,B).

Although average wind speed is no greater than in temperate latitudes, gales are common. Wind is of most importance in winter since it removes snow from convex surfaces and deposits it in depressions, and the depth of accumulation is crucial in both arctic and extratropical alpine tundras. Where the snow mantle is kept thin, few species can endure exposure to desiccation, ice-particle blast, and chilling. On the other hand where snow accumulation is more than barely enough to afford good protection, lateness

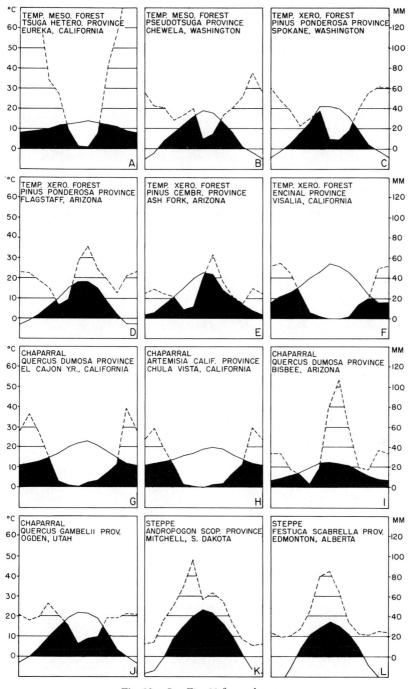

Fig. 12. See Fig. 11 for explanation.

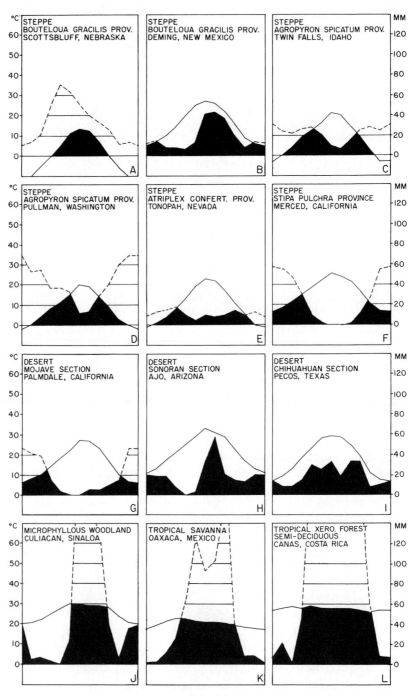

Fig. 13.   See Fig. 11 for explanation.

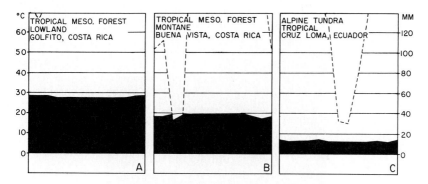

Fig. 14. See Fig. 11 for explanation.

of melting in summer shortens the growing season and places a different kind of restriction on vegetation composition. Thus, plant communities commonly form concentric belts about snow accumulations (i.e., "snow patches" or "snow beds"), each belt reflecting a different degree of shortening of the growing season (see Fig. 15).

"Physiologic drought" that might result from the limited availability of water in cold soil has long been thought to account for the frequency of tundra xeromorphy, with dwarfness and frequency of winter injury on exposed ridges cited in support of this interpretation. However, since transpiration is an endothermic process, its cooling effect on leaves in an environment where low temperatures are normally critical may be as important a cause of dwarfness as the drying effect of wind. Furthermore, xeromorphy may be of value in the endurance of abrasion by wind-driven ice or soil particles, or it may be mainly a response to the low level of fertility. Low temperatures by themselves appear to be unimportant, for all the plant organs freeze in winter, and hard frosts during summers seldom affect more than floral organs.

The soil horizons in which the plants are rooted is usually frozen for at least nine months. Those tongues of forest that extend far into the arctic tundra along valleys may reflect either a measure of protection from soil freezing as a result of earlier snow accumulation in valleys, or they may reflect a critical reduction in the velocity of drying winds in winter.

In continental climates permafrost in zonal soils is usually encountered below about 0.3–2.0 m. By intercepting most of the solar radiation in summers, vegetation prevents deep thawing, so when raw substrates are colonized permafrost rises nearer the surface. In autumn the thawed or "active" layer thins by freezing, which progresses upward from the permafrost as well as downward from the surface. As ice crystals expand, the unfrozen trapped material is subjected to strong pressure. In places mud is squeezed out through weak spots in the crust. Elsewhere the crust is lifted in mounds of varying sizes and shapes. This phenomenon is common even

Fig. 15. Vegetation zonation related to snow depth in Swedish Lapland. *Salix herbacea* with cryptogans dominate the bottom ²/₅ of the picture, where snow persists longest in winter. The herbaceous belt dominated by *Deschampsia flexuosa* is next upslope, growing where snow melts away earlier. A *Vaccinium myrtillus* belt follows where the snow cover is thin. On the ridge crest under the feet of the men *Loiseleuria procumbens* and *Empetrum hemaphroditum* occur where snow accumulates only among the plant shoots.

where the vegetation remains undisturbed, but where the protective cover is destroyed the soil can thaw more deeply, and the wider annual cycles of change between liquid and solid states of water magnifies this frost churning of soils. Substrates are thus unstable over extensive areas, such activity preventing the development of horizons in soil profiles, and producing stone nets and stripes, mounds ("pingos" and "frost-boils"), solifluction lobes (Fig. 16), ice-wedge cracks, etc. Seral stages, microseres, and retrogressive succession therefore characterizes much if not most of the tundra landscape. Any destruction of the plant cover by traffic, prospecting, etc., sets off soil churning and so destroys the productivity of the affected area for an indefinite period. In continental areas, even where frost-churning and solifluction are not aggrevated by man's activities, these phenomena tend to be more important as agents remodeling the land surface than are water erosion and deposition, which are the most effective agents at lower latitudes.

Fig. 16. Solifluction lobe lifting tundra sward as it advances periodically. Kopparason, Sweden.

In oceanic climates where summers are longer but cooler, the winters milder, and precipitation higher, permafrost is typically absent.

Chemical weathering of parent materials is so slow that there is negligible clay formation. The mineral material remains coarse-textured and stony, with silt better represented than clay-sized particles.

Low temperatures and acidity in the range of pH 3–5 retard litter decay. As a consequence much fertility is immobilized in a 3–30-cm layer of litter and duff that accumulates wherever there is a good cover of vegetation on stable soil. The low fertility of tundra soils is shown by the circle of lush vegetation that develops around isolated boulders used as look-out perches by birds and mammals that leave their wastes there, and by the lush vegetation of nesting sites of water birds. Village sites long abandoned have especially lush vegetation owing to the refuse of bone and shell middens plus human wastes. The contrast between floras of calcareous and siliceous parent materials is also very strong, and since acid soils are more normal to cold, wet climates, the stable vegetation of calcereous soils must be considered as edaphic climaxes.

The combination of short, cool summers plus low soil fertility makes for low primary productivity in tundra ecosystems (Table 3).

Plant life.[351]   Only about 900 species of vascular plants (230 genera, 66 families) are involved in the entire circumpolar tundra belt of the Northern Hemisphere, with the families Compositae, Cyperaceae, Gramineae, Cruciferae, and Caryophyllaceae best represented. It has been said that

TABLE 3

Estimated Net Primary Productivity of Major Vegetation Types, as Adapted from Lieth[257]

Tropical Mesophytic Forest	1000–3500[a]
Tropical Xerophytic Forest	600–3500
Temperate Mesophytic Forest	400–2500
Subarctic Forest	200–1500
Chaparral	250–1500
Steppe	100–1500
Tundra	100–400
Desert	0–250
Freshwater	
Swamp and marsh	800–4000
Ponds, lakes, and streams	100–1500
Oceans	
Estuaries and reefs	500–4000
Continental shelf	200–600
Upwelling zones	400–600
Open ocean	2–400
Cultivated land	100–4000

[a] Grams per square meters per year.

nearly one-third of this flora is alien.[260] Bryophytes, lichens, algae, and fungi each would seem to provide as many species as the vasculares. Ferns are rare, and liverworts few.

Floristic affinities between Alaska and Asia are closer than affinities between Greenland and Scandinavia, indicating that either land connections between the latter two were broken long ago, or have been less continuous. This flora has been well collected, but in North America little work has been done on the communities and their ecology.[47,94,173,222,321,380] Good examples of the synecologic possibilities are illustrated by Scandinavian work.[93,154]

Dwarf shrubs (chamaephytes) are especially well represented in arctic tundra. Annuals are extremely few, probably because they cannot construct sufficient photosynthetic apparatus to make use of the short summer until much of it has passed. Then too, should early or late frosts prevent seed set or maturation, this would be disastrous.

Many of the species are evergreen, or at least have the lower leaves tending to overwinter. Evergreenness is probably significant as an adaptation permitting the longest use of materials used in leaf construction, together with the advantage of being able to function during short periods of warmth at the beginning and end of winter.[377] Despite this prevalence of evergreenness, the vegetation as a whole has a drab yellow–green color as a result of standing dead litter from the preceding season that decays very slowly.

As pointed out earlier, low stature enables plants to benefit from the protection afforded by snow cover in winter, and poorly developed bud scales do nothing to weaken this interpretation. Low stature also allows plants to take advantage of the warmth at the soil–air interface. At this level both tissues and soil thaw most frequently, and minimal temperatures for photosynthesis, which are approximately 0°C in these plants, are attained most frequently.

Most upland vasculares of arctic tundra have small, entire, firm-textured or hairy leaves, reduced stems, and enlarged subterranean organs devoted to carbohydrate storage. Those with overwintering foliage store moderate to large quantities of lipids above ground. The stomata remain open all day so that these plants accumulate dry matter rapidly, and with the general wetness of the environment, open stomata pose no problem from the standpoint of transpiration stress, except in the dry far-northern belt. Net primary productivity in closed herbaceous communities is usually around 100 gm/ m²/hr, but is several times greater than this in the shrub communities (Table 3).

The plant body is highly efficient in its construction, for the ratio of shoot to under ground organs is about 1:5, and with very little stem tissue above ground (Fig. 17), most cells of the plant are directly involved with either the manufacture or storage of food.[225] Despite the conspicuous lack of bud scales, persistent old leaves protect the growing tips of the stems. By the end of summer the leaf primoidea of the next season are well developed in many

Fig. 17. In tundra communities *Salix* commonly has a prostrate if not subterranean stem, with only short twigs rising above the soil surface. Glacier National Park, Montana.

species, and by then food reserves have been restored, with a result that in spring the new shoots expand rapidly as the snow melts, and so can make the most of a short summer.

The osmotic potential of tundra plants is relatively high in summer, but declines at the onset of winter as free water is mostly eliminated from tissues. This, along with small cell size, is of undoubted importance in the ability of the tissues to withstand freezing.

In about half the species the flower buds that develop in autumn while fruits and seeds are ripening, may be viewed as an overlapping of reproductive cycles that helps offset the limitations of a short summer. In Greenland the flowering season is from late May to mid-August, attaining a maximum in early July.[377]

Cross pollination in dicots is accomplished mainly by insects, with graminoids depending on wind, but nearly all species are self-fertile.[351] Wind dissemination prevails, especially by protracted gales that scatter disseminules over the snow surface in winter.

In arctic tundra more than anywhere else, there have developed reproductive strategies that bypass the sexual process. *Saxifraga flagellaris* produces stolons which root at their tips. In *Polygonum viviparum* some of the flowers are replaced by loosely attached bulblets that can take root when they fall from the inflorescence. Ovules in *Taraxacum* can develop into viable seeds without being fertilized. Reproductive characters such as these seem of advantage to the extent that the procedure is more certain of success and involves less time than is required by pollination, fertilization, embryo development, and seed maturation. Also they may be interpreted as mechanisms that have survival value by perpetuating any advantageous combinations of genetic characters that might arise by accident.

Polyploidy, rising to 85% of the flora in places, is far more prevalent in the arctic tundra than at lower latitudes. However, we do not know if polyploidy has been favored by conferring a superior tolerance of cold climate (an old suggestion that has been questioned), or if it has been a response to unstable climate and soil during deglaciation, or if it is simply a result of arctic climate favoring herbaceous perennials which happen to be mostly polyploid.[161] Also, since most polyploids are of hybrid origin, they have more genetic diversity, hence relatively large variability, and this confers an advantage.[221]

If attention is confined to zonal soils, tundra physiognomy varies with distance beyond the forest edge. Typically there is first a belt of rather dense vegetation in which a mixture of deciduous and evergreen shrubs mostly less than a meter in height are conspicuous if not dominant, with the ground beneath usually moss-covered. The major taxa found in the different geographic areas include *Arctostaphylos arctica, Betula, Cassiope tetragona, Diapensia lapponica, Empetrum, Ledum decumbens, Loiseleuria procumbens, Phyllodoce coerulea, Rhododendron lapponicum, Salix, Vaccinium uliginosum,* and *V. vitis-idaea.* Where the soils are rather wet, *Betula nana* and *Salix* tend to prevail over the other shrubs. Species diversity is probably higher in this shrub belt than elsewhere in tundra. Such scrub extends farthest northward in slight depressions and other places where snow provides good protection in winter. Along the southern margin it becomes a major component of the broad forest–tundra ecotone, alternating with groves of trees which occupy more and more of the land southward. It characterizes the coastal margin of Greenland.

In slightly colder climate there tends to be a belt of mainly closed communities no more than about 3 dm in height, which are dominated chiefly by perennial graminoids and forbs, but include prostrate *Salix* and a few other shrubs. Some common graminoid genera here are *Arctagrostis, Carex, Deschampsia, Festuca, Poa, Kobresia,* and *Luzula.* These are intermixed with dicots in *Artemisia, Astragalus, Cerastium, Draba, Epilobium, Papaver, Polygonum, Potentilla, Saxifraga,* and *Stellaria.* Mosses and lichens form a rather poorly developed layer beneath these vasculares. On south-facing slopes where permafrost is deeper, species diversity is the higher.

The soils of this dense turf tend to be less acid than under the shrub belt. In continental parts of Alaska this belt with its Arctic Brown Soils is rather poorly represented since most of the landscape is poorly drained. In oceanic climates of the Bering Strait area Subpolar Meadow Soils are characteristic.

In the third major tundra belt, extending to the edge of perpetual snow, lichens, especially species of *Alectoria, Cetraria, Cladonia,* and *Thamniola,* provide the most continuous layer. Vasculares of very low stature and mainly of the cushion or rosette form are so widely scattered that the landscape has a rather barren aspect, with the greater phytomass concentrated in the wetter sites. *Draba, Dryas, Papaver radicatum, Salix arctica,* and *Saxifraga oppositifolia* are widespread representatives of this impoverished flora. Vegetation of this type is characteristic of places where the mean monthly temperature as recorded in a standard instrument shelter does not rise above 3.5°C.[261] However, the value of standard weather station data in understanding the ecology of tundra plants is limited by the fact that temperatures as recorded at 1.5 m above the ground are considerably lower than those at the level of ground-hugging plants.

Although the soil here is relatively dry, there is sufficient moisture for the development of stone nets and other types of patterned ground. The Polar Desert Soil associated with this extremely sparse vegetation has virtually no profile development or humus, has a desert pavement resulting from deflation, and is circumneutral. Owing to the dryness local spots may have a thin salt crust on the surface,[389] especially in the interior of northern Greenland. Thus, there is some justification for calling this part of arctic tundra a "polar desert."

In regions of perpetual snow, highly specialized types of algae in a number of genera provide organic compounds that support fungi, rotifers, and tiny worms. Depending on the dominant species, the algae render the snow faintly pink, yellow, or green.[80]

All of the above-described vegetation belts interfinger, with each type of vascular vegetation extending farthest poleward in the protection of slight depressions, then extending farthest equatorward on convex topography where the microclimate is relatively severe.

Despite the outstanding floristic poverty of arctic tundra, the ecosystems are well diversified. Ecosystems widely referred to as fell-fields have Lithosols on which *Dryas, Saxifraga oppositifolia, Silene acaulis* and other mat plants grow in open stands with a few species of *Carex* and *Juncus*. Much of the surface here consists of lichen-encrusted stones.

Streambanks are commonly lined with tall shrubs belonging to *Alnus crispa* or *Salix*. Seasonal marshes and fens with Tundra or Sod Glei Soils occur in extensive tracts on coastal plains or other flat topography. Characteristic plants here include *Arctophila fulva, Arctagrostis, Calamagrostis, Carex aquatilis, Dupontia fischeri, Eriophorum,* and *Juncus*, these communities often with a well-developed moss layer. Bogs have hummocky surfaces of *Sphagnum* usually dotted with shrubs such as *Rubus chamaemorus* or ericads including *Andromeda polifolia, Ledum decumbens,* and *Vaccinium uliginosum*. Ponds are very abundant, many formed by the local thawing of permafrost, then later disappearing by a reversal of the same process. Common hydrophytes growing in the water are *Arctophila, Caltha palustris, Carex, Equisetum, Hippurus vulgaris, Menyanthes trifoliata, Potamogeton, Potentilla palustris,* and *Sparganium*. Tidal marshlands along coastlines support the widespread grass *Puccinellia phryganoides. Elymus arenarius* is the grass most common on dunes.

In places where snow accumulates deeply and shortens the growing season there are concentric belts of vegetation about a central area that may be bare or supports only *Salix herbacea* and cryptogams, owing to its brief exposure to the sun each summer.

Animal life and land use.   Characteristic herbivores of the tundra that are permanent residents include musk ox, caribou, arctic hare, ground squirrel, lemming, ptarmigan, snow bunting, lapland longspur, with carnivores represented by polar bear, grizzly bear, wolf, arctic fox, weasel, snowy owl, falcon, raven, and jaeger. Lemmings, active the year around, consume much herbage and churn the soil with their digging, especially in the frequent years when they have a population eruption. Foxes add to soil disturbance by digging out rodents. Caribou move about constantly and so do negligible damage by grazing, but in their trails and other minor areas of concentration plant life is destroyed by trampling. Some caribou populations retreat southward into the forest during winter, but others remain on the tundra, subsisting mainly on lichen vegetation that remains available on windswept ridges. Since lichens are long lived they accumulate airborne DDT progressively, then pass this on to caribou where it becomes concentrated in the tissues, with eskimos in turn accumulating still higher concentrations when they eat caribou.

Vast numbers of waterfowl and shore birds which migrate to the fens, marshes, and ponds of arctic tundra to breed each summer, later provide much of the hunting enjoyed by the inhabitants of temperate latitudes when the birds migrate southward in autumn. While on the breeding grounds, ducks and geese graze, compact, and fertilize the vegetation. A major dam that has been proposed to flood a large portion of the Yukon Valley would be disastrous for this waterfowl and other wildlife, whereas the value of so much hydroelectric power generated in a wilderness has been questioned.

The eskimo lived entirely in the arctic tundra, with some subcultures based on hunting sea mammals and fish, and others dependent primarily on caribou. Although they tore up woody roots and stems for fuel, their influence on terrestrial vegetation was negligible in view of the thin populations such a harsh environment supported and their use of animal oils for lighting.

No commercial agriculture is possible in the arctic tundra, but in favorable locations crops such as white potato, turnip, cabbage, rutabaga, celery, lettuce, and radish can be grown in home gardens. Fertilizer is essential.

Management of semidomesticated reindeer, the close Eurasian relative of the American caribou, continues to be the best means of utilizing the natural vegetation resource indefinitely, as is well demonstrated by sound management practices operating in northwestern Europe. A special consideration in this type of land use is the slow growth of lichens. Up to three decades are required for regrowth of a plant that has been grazed to near the soil surface. In Alaska there has recently developed some interest in domesticating musk ox for both meat and wool.

The major disturbance of arctic tundra by modern man has resulted from exploration for oil and the construction of a pipeline for its transmission.[43] Recent activity of this nature, combined with a lesser amount of disturbance during the war in the Pacific, has caused more damage than millennia of eskimo occupation. Massive campsites littered with debris, extensive and broad trails marked by destroyed vegetation, drained lakes, etc., have wrought changes in drainage, and initiated frost churning of the soil, that will persist for centuries.

Extratropical Alpine Tundras

Although alpine and arctic tundras are essentially continuous, there are substantial differences in the environment of modal localities.[34,40] Alpine tundras have a much wider range of temperature than does the arctic tundra. Although the soil surface in the mountains is subject to freezing in winter, permafrost is relatively rare, and where it occurs it is relatively deep and unimportant ecologically. Frost churning of the soil is consequently negligible, although "fossil" stone nets and stripes bespeak its importance during

Wisconsin glaciation. Wind velocity is generally higher in the alpine than arctic tundra. Precipitation may be higher or lower (Fig. 11C,D). Daylength during the growing season shortens progressively in an equatorward direction, and ultraviolet radiation at high altitudes is greater than in arctic latitudes. Although CO_2 tension is lower, a higher diffusion rate keeps this from being a limitation to photosynthesis.

About half of the species of arctic tundra extend for varying distances southward, despite these major differences in arctic and alpine environments, emphasizing the overriding importance of heat-deficiency in both. The physiognomy and ecology of alpine vegetation is very similar to that of the arctic tundra, but it has the higher species diversity where the areas are relatively large. Arctic representatives in the alpine floras are either ecotypically or ecoclinally distinct. In contrast with conditions in the north, diploidy is more prevalent than polyploidy.

In Europe it is generally recognized that the use of wood for fuel by herdsmen while they tended livestock in alpine meadows has lowered tree lines perhaps as much as 200 m. There is no evidence of this in North America. On the contrary, owing to cooling, the weathered remains of trees above present tree lines show that in the Cordillera of the west central North America the tree line has receded at least 150 m in post-Hypsithermal time.[248] Whereas arctic tundra forms an ecotone with forest that is hundreds of kilometers broad, alpine ecotones are surprisingly abrupt, with tall trees forming closed forest to within 100 m or so of the elevation of the highest stunted grove.

Appalachian Section. [12,41,210]   Alpine tundra extended more than 300 km farther south down the Appalachian Plateau during Wisconsin glaciation than it is found on the mountain summits today.[283] Later, as temperatures rose, fragments receded to high altitude in the few mountains available and became surrounded by subalpine forest. These fragments occur in New York, New Hampshire, and Maine, with other hills rising above timberline on the Gaspé Peninsula and in Labrador. Even though timberline is lower here than in the western Cordillera at comparable latitudes (Fig. 10), the mountains are so small that few summits are high enough to be too heat-deficient for trees. In contrast with arctic timberline, the forest margin here has been interpreted as receding.[163]

The shrub belt of this alpine tundra includes *Arctostaphylos alpina, Betula glandulosa, Diapensia lapponica, Empetrum nigrum, Loiseleuria procumbens, Phyllodoce coerulea, Vaccinium uliginosum,* and *V. vitis-idaea,* all of which are represented in arctic tundra as well. At least 75% of the flora is shared with the arctic belt. *Arenaria groenlandica, Carex bigelowii, Juncus trifidus,* and *Potentilla tridentata* are major dominants of the herbaceous communities. Lichens are much better represented in these mountains than in the Cordillera, reflecting the abundance of fog and high relative humidity.

Most of this vegetation is on Alpine Turf Soils, but Podzol occurs at least locally.[44]

Rivulets are bordered by shrubby *Salix*, with bogs and fell-fields adding further to landscape diversity. *Abies balsamea* and *Picea mariana* are the principal trees at the lower ecotone (Fig. 18).

Rocky Mountain Section. [86,226,361] Rocky Mountain tundra has been studied chiefly in the central Rockies, where there are large areas above climatic tree limits, despite the relatively high elevation of the latter (Fig. 10). The leading genera here are much the same as in the arctic tundra, and a number of species are shared with the arctic. In southern Montana 50% are shared, in Colorado 37% are arctic, and in Arizona at the southern limit of the Section 30% occur in the arctic.

In the far-northern Rockies *Alnus*, *Betula glandulosa*, and *Salix* dominate a well-developed shrub belt in the lower alpine belt of the interior mountains. Near the Pacific Ocean at this high latitude a low scrub dominated by *Luetkea pectinata* and *Selaginella* is common just above forest limits. To the south in the vicinity of the Canadian border the principal shrubs are the evergreen *Cassiope mertensiana*, *Empetrum nigrum*, *Phyllodoce empetriformis*, *P. glanduliflora*, and deciduous species of *Vaccinium*. In northern Wyoming only shrubby *Salix* with small amounts of *Phyllodoce empetriformis* and *Potentilla fruticosa* represent the shrub belt, and farther south this element of physiognomy is absent.

Above the shrub belt, or tree line, zonal soils support a dense, but low and meadowlike sward (Fig. 19) dominated by graminoids in *Carex*, *Deschamp-*

Fig. 18. Timberline ecotone with *Picea mariana* krummholz, Mt. Katahdin, Maine.

Fig. 19.   Climax alpine meadow. Medicine Bow Mountains, Wyoming.

sia, Festuca, Kobresia, Poa, and *Trisetum,* interspersed with such forbs as *Geum rossii* and *Polygonum bistortoides. Koenigia islandica* is the only annual in this flora. Soils associated with such vegetation have been classified as Alpine meadow or Alpine Turf.[336]

Spagnum bogs, and fens with *Carex, Eleocharis, Eriophorum, Menyanthes,* etc. are common. Fell-fields having *Arenaria, Dryas, Luzula spicata, Selaginella, Sibbaldia procumbens,* and *Silene acaulis,* occupy large areas (Fig. 20). Boulder fields, consisting of large angular frost-riven blocks left as a jumbled mass on the surface by Pleistocene climate, support little more than crustose lichens (Fig. 21).

Those places where snowbanks persist into summer are marked by special communities in which *Carex nigricans, Juncus drummondii, Erythronium grandiflorum, Leutkea pectinata,* and *Valeriana sitchensis* are representative plants. As at high altitudes, snowbanks become tinted by algal communities in summer.

Typical vertebrates include bighorn sheep, mountain goat, marmot, pika and ptarmigan, which are preyed upon by puma, grizzly bears and (formerly) by wolves. Herds of wapiti often move up onto alpine tundra in summer.

Cascade-Sierra Section.[66]   Alpine environment was late to develop in the Sierra–Cascade mountain axis since these mountains did not rise above tree limits until late in Pliocene time. Furthermore, proximity to the ocean makes their summer-dry climate far less suited to the requirements of most arctic tundra plants than are the climates of the Rockies and Appalachians, so

Fig. 20.   Alpine fell-field, an edaphic climax dominated by *Silene acaulis*, *Trisetum spicatum*, *Sieversia turbinella*, and *Selaginella*. Beartooth Mountains, Montana.

a boreal source of adapted species has been very limited. These circumstances, coupled with the numerous discontinuities in the alpine environment, resulted in the development of an alpine vegetation drawn mostly from local floras at lower elevation. Less than 20% of the alpine flora of the Sierras is also represented in the arctic tundra.

In the Cascades *Cassiope*, *Empetrum nigrum*, *Phyllodoce*, and *Vaccinium deliciosum* are among the most conspicuous shrubs in the lower margin of

Fig. 21.   Boulder fields, areas covered with frost-riven rock, are common in alpine environments.

the alpine vegetation, where the soils are Podzols. In contrast with the interior mountains where graminoids dominate over forbs, on Alpine Turf Soils in the Cascades the major dominants are forbs such as *Arenaria, Aster, Erigeron, Lupinus lepidus, Pedicularis contorta, Phacelia, Phlox diffusa*, with *Carex spectabilis* the most outstanding graminoid. *Carex nigricans* comes into prominence on snowbeds.

In the Sierras too, graminoids play a minor role in the mountain meadows. Southward in these mountains rainfall becomes unusually low for alpine environment, and there annuals are well represented, these having been derived from the desert flora of the contiguous basal plain.

Until recently sheep grazing was the chief force bringing about ecosystem deterioration in both the Rockies and the Sierra–Cascade Mountains, but recently recreation use, especially the incursion of off-road vehicles, has burgeoned to impose an even greater stress upon these ecosystems.

Mexican Section.   Although the few volcanic peaks that rise above timberline in Mexico and Central America are well within the astronomic tropics, their alpine vegetation will be considered here since the flora consists of genera of northern affinity.[26] However, almost none of the plants are closely related to the alpine tundra of the Rockies, and affinities with the alpine flora of South America are equally weak. *Phleum alpinum*, found in northern Mexico, and *Tristeum spicatum* in central Mexico, appear to be the only species shared with the North American arctic–alpine flora. The high peaks of central Mexico, Ixtaccihuatl and Popocatopetl, provide the terminus for three genera of Andean affinity (*Acaena, Colobanthus, Pernettya*), with two species in this category extending as far north as Nuevo Leon.

Except for a limited representation of *Vaccinium*, shrubs are almost wholly lacking in the alpine vegetation of this Section, and *Carex* does not play the important role here that it plays in alpine floras to the north. The prevailing vegetation is meadowlike, consisting of genera that are familiar on the lowlands of temperate North America: *Agrostis, Androsace, Arenaria, Bidens, Calamagrostis, Castilleja, Cerastium, Draba, Festuca, Geranium, Hieraceum, Muhlenbergia, Potentilla, Senecio*, etc.

Antarctic Tundra[196]

The extreme cold of Antarctica has imposed severe limitations on the development of tundra vegetation. In the interior, temperatures may drop as low as −84°C, and since there is neither rain nor snowmelt over most of the land, environments are dry as well as cold.

The known flora of that continent consists of two species of vasculares (*Deschampsia antarctica* and *Colobanthus crassifolius*) at the warmest edge, 4 species of *Agaricaceae*, 70 species of *Bryophyta*, 200 species of algae, and at least 400 species of lichens.

The few treeless islands scattered over the Antarctic Ocean collectively have a vascular flora of only about 85 species—less than one-tenth of the arctic tundra flora.

THE SUBARCTIC–SUBALPINE FOREST REGION

The Subarctic–Subalpine Forest Region includes extensive areas of species-poor forest dominated by various combinations of *Abies*, *Larix*, *Picea*, and white-barked *Betula* or *Populus*, forming a belt adjacent to arctic and alpine tundras of the Northern Hemisphere. It does not include all needle-leaved coniferous forest vegetation, especially forests characterized by *Juniperus*, *Pseudotsuga*, *Sequoia*, *Thuja*, many taxa in *Chamaecyparis* and *Pinus*, and some in *Abies* and *Picea*. Thus, it is interpreted as including only a part of that morphologic category which physiognomically based classifications group together as needle-leaved or narrow-sclerophyll forests. Physiognomy by itself is obviously unreliable as an indication of ecologic relations, which are given more weight here.

Taiga is a Russian word that appears to have been used originally for the broad ecotone between tundra and subarctic forest in Eurasia. Its use has gradually been extended to include subarctic forests in both Eurasia and North America, and to a limited extent it has been applied to subalpine forest in the latter continent.

Origin and Distribution

Angiosperms did not crowd conifers completely out of existence as they overran Cretaceous landscapes, for along the northern border of Laurasian land fragments and on the summits of mountains conifers retained at least a subordinate role. Nearly all fossil assemblages of Eocene and other early Cenozoic deposits found in extratropical North America include such genera as *Chamaecyparis*, *Metasequoia*, *Sequoia*, *Taxodium*, *Thuja*, and *Tsuga* which were growing intermingled with trees (*Acer*, *Fagus*, *Liquidambar*, *Tilia*, *Ulmus*, etc.) indicating temperate rather than subarctic environment. But on Grinnell Land, only 8.5° from the North Pole, an impoverished flora has been described[32] in which the chief genera (*Betula*, *Corylus*, *Picea*, *Pinus*, and *Salix*) suggest an Eocene arboreal forest ancestral to the present subarctic forest.

However, even Beringia was occupied by temperate forest until at least late Pliocene time,[175] and although Cenozoic cooling then made the climate suitable for floristic exchanges between the subarctic forests of Asia and North America, the contact was prevented by a submergence of the former land bridge.[413] Not long thereafter cooling resulted in tundra replacing subarctic forest in the Beringia area, and floristic exchange between subarctic forests has ever since been precluded.

As Cenozoic cooling forced the subarctic flora ever farther southward, peninsular extensions penetrated farthest along the cool summits of mountains as these became available in temperate latitudes. Also as the temperatures dropped, some preadapted species and ecotypes of the temperate floras which had clothed the mountains that were becoming chilled, were able to maintain their positions and mingle with immigrants from the north.

The derivation of most major elements of subalpine forests at low latitudes from the subarctic flora is clearly indicated by the facts that *Abies fraseri* of the Appalachians and *A. lasiocarpa* of the Rockies are scarcely more than subspecies of the subarctic *A. balsamea*, with *Picea rubens* of the Appalachians and *P. engelmannii* of the Rockies equally close to the subarctic *P. glauca*. *Populus tremuloides* is scarcely differentiated at the subspecies level as it extends southward in each of the three major mountain systems.

Since the principal centers of Pleistocene ice accumulation were located well within the subarctic belt, much if not all of that area was preempted by ice repeatedly. During Wisconsin glaciation, at least, subarctic forest was eliminated from all the area which it probably occupied previously, so that those species which survived did so by migrating to temperate latitudes where they infiltrated among members of the temperate mesophytic forest, or invaded the previously arid plain east of the Rockies as lowered temperatures increased the effectivity of precipitation. The ice advanced too rapidly for a distinct belt of subarctic forest to maintain dominance in front of it, for in places the ice actually pushed over living trees, and in midcontinent it advanced directly onto steppe. Only as the ice advance slowed to a stop did trees have time enough to spread over the formerly arid plain. This is not to imply continuous forest, but at least the most mesophytic habitats on the plain came to support trees. Macrofossils of *Abies*, *Picea*, and *Thuja* have been found in Louisiana, and fossil pollens show that the subarctic trees infiltrated into Georgia and Texas as well. In the western mountains, cooling had the effect of displacing altitudinal vegetation belts downward to a distance estimated as about 1000 m.

Post-Wisconsin northward migration of the subarctic flora as temperatures rose was much more rapid than the southward infiltration into temperate zone vegetation had been. Numerous relic colonies were left stranded in habitats of compensation south of the present limits of the subarctic forest belt. Relictual stands of *Abies balsamea* have persisted on steep north-facing slopes in Iowa[77] and southeastern Minnesota[299]; *Betula papyrifera* became stranded at many places in Nebraska and left a hybrid swarm in Colorado[148]; disjunct populations of *Picea glauca* occur in the Black Hills of South Dakota with *P. glauca* × *P. engelmannii* hybrids in the Big Horn Mountains of Wyoming and the Cypress Hills of Saskatchewan.

Northward migration continued to the culmination of Hypsithermal warmth, followed by southward retraction during subsequent cooling. There is good evidence that in California alpine timberline has shifted downward at least 150–200 m since the culmination of Hypsithermal warmth.[244] Possibly some mountain summits now having climates suitable for subalpine forest lack such vegetation owing to its having been "squeezed off" during the warm period.

During the last half century the northern border of the subarctic forest has again been advancing poleward both in Alaska[161] and Europe. Although there has been no detectable change in mean annual temperatures in the tropics during recorded weather history, between 1920 and 1960 temperatures rose 2°C in the United States and several times as much in Spitzbergen.[117] In temperate latitudes this has not been sufficient to provide clear evidence of upward migration.

Tree species derived from the subarctic flora have reached no farther south than southern Arizona (*Picea engelmannii*) and central Mexico (*Populus tremuloides*). Thus, as subalpine environments developed on mountains in Mexico and Central America, subalpine forest dominants had to be drawn from the floras of the midmountain slopes, and there *Pinus* proved to be more adaptable than any other genus. A homologous belt of pine-dominated forest likewise occurs in the tropical mountains of Indochina.[412]

In Europe the stresses of Pleistocene glaciation were so severe that all ecotypes of *Abies* and *Larix* which were adapted to subarctic forest environment became extinct, so that only the paper-barked *Betula tortuosa* and *Pinus sylvestris* have been left to dominate the most northern part of the subarctic forest belt, with *Picea abies* joining them in the southern part of the belt.

In North America the major segment of the Subarctic–Subalpine Forest Region forms a broad belt that stretches from the Atlantic coast of Canada to near the Pacific coast of Alaska and British Columbia. In the Appalachian Mountains and in the Cordillera of western North America, subalpine forests closely similar in environment, if not in floristics, occur wherever the mountains attain sufficient height. In Mexico the somewhat ecologically related subalpine forest extends well below the Tropic of Cancer.

Subarctic–subalpine forest is even more extensive in Eurasia, and on that continent too it is represented on all the major mountain masses from the Pyrenees to the Himalayas. No corresponding vegetation unit has been recognized in the Southern Hemisphere.

Environment

Winter temperatures may be lower in the subarctic forest than in the Arctic Tundra Region, since the latter borders the Arctic Ocean which has a damping influence on temperature extremes (Compare Fig. 11B, E). The lowest temperature ever recorded at a permanent weather station is −71.1°C at Oymyakon, Siberia, a station surrounded by forest. Although the mean temperature of the warmest month in such continental climates may rise well above 10°C, frost is common even in this month.

As pointed out earlier, the climatic characteristic most closely related to the tundra–forest ecotone is the level of midsummer temperatures (see Fig. 11A–H). The arctic ecotone correlates well with the average position of the Arctic Front in summer, when the Cold Polar air mass has contracted the most.

At different latitudes the species of trees reaching the alpine limits of forest have different ecologic amplitudes, but the facts that alpine timberlines have a symmetrical and predictable pattern in relation to latitude and altitude, and that this pattern correlates well with the warmest-month isotherm of 10°C, show that there is a fundamental physiologic limitation to the tree life form that overshadows these species' differences. It has been hypothesized that cold timberlines represent a critical level in the diminishing heat supply beyond which a plant cannot afford so many nonphotosynthesizing cells as in a normal tree, and dwarfing of the massive woody organs results. The importance of temperature is also reflected in the higher elevation of upper timberlines on equatorward slopes in comparison with poleward slopes, and in the Massenerhebung effect (higher timberlines on more massive mountains at a given latitude), both phenomena being correlated with elevated isotherms.[100]

Locally, upper timberlines in the mountains are determined by many factors other than heat supply. Timberline is often relatively low on a small isolated mountain that is exposed to strong winds which have a desiccating, hence dwarfing, effect. Where there is a heavy snowfall that accumulates to great depths in convex topography, the snow-free season may be shortened too much for tree seedlings to complete necessary development during their first summer. In such areas trees extend to highest elevations on ridges or knobs where snow accumulates least and melts away first. Vast slopes of loose talus that extend from alpine tundra far down the mountains are perhaps the most common cause of depressed timberlines. In the Alps where herdsmen watch their livestock during the summer grazing season, it is suspected that over the centuries their use of wood for fuel has lowered timberline appreciably.

Most trees show marked reduction in the frequency of seed production as they approach cold timberlines. Although the summers seldom have sufficient heat to favor the initiation of flower buds, the rarity of seed years is compensated by the longevity of the slow-growing trees. Also vegetative reproduction by layering, favored by persisting lower branches and the heavy weight of snow over long periods, helps maintain populations in certain trees. The populations of some species extend up mountains as far as climate permits only by virtue of seeds brought up from the forests below on updrafts.

In subarctic forests precipitation falls mostly in summers, but in the subalpine forests of the Cascade Mountains it falls mostly in winter, and in the Appalachians there is a more equitable distribution throughout the year. Regardless of seasonal distribution, the amount of precipitation is always adequate to keep the soil from drying more than a few centimeters deep during the short, cool summers.

Differences in climate (daylength, summer temperatures) are great enough between the subarctic Province and the subalpine Provinces, that there is little overlap in the ranges of the dominant trees. On the other hand, since forest undergrowth is less closely related to macroclimate, there is an appreciable sharing of shrubs and herbs between subarctic and subalpine Provinces.

The *Picea glauca* Province[208,249,298]

Zonal Vegetation

Picea glauca (Figs. 9 and 22) is the ubiquitous dominant of climatic climaxes in this Province, but *Betula papyrifera*, an aggressive invader of burned areas which sprouts readily from the root summit when the trunk is killed, persists indefinitely but in reduced numbers in the stabilized communities. Across the northern border of the Province, and toward the upper altitudinal limits southward, *Picea mariana* becomes a climax dominant on well-drained upland loams.

The undergrowth of these forests consists of a number of widespread herbs and shrubs (*Clintonia, Coptis, Cornus canadensis, Goodyera repens, Gymnocarpium dryopteris, Linnaea borealis, Listera cordata, Lycopodium annotinum, Maianthemum, Mitella nuda, Pyrola uniflora, Rosa acicularis, Trientalis, Virburnum edule*), with moses, especially *Hylocomium splendens, Hypnum schreberi,* and *Ptilium cristi-castrensis,* often forming a continuous ground cover in certain associations otherwise poor in undergrowth.

Two east–west Sections may be recognized. Eastward from the Rockies *Abies balsamea* becomes an increasingly more important member of the climaxes at low latitudes, and since it is more shade-tolerant than *Picea glauca* it tends to replace the latter. However, owing to its relatively short life span and the frequence of fire, it seldom completes replacement between holocausts. In addition to the ubiquitous undergrowth species listed above, *Acer spicatum, Gaultheria procumbens, Oxalis montana, Trillium,* and *Vaccinium pennsylvanicum* are well represented in only the eastern Section.

The western or Cordilleran Section lacks *Abies balsamea.* Special additions to the climatic climax undergrowth here include *Lupinus arcticus* and *Shepherdia canadensis.*

Fig. 22. *Picea glauca* and *Betula papyrifera*, 97 km south of Ft. Nelson, British Columbia.

Acids formed when conifer litter decays promote the loss of soil bases by leaching, and the low nutrient content of the litter does little to offset the resultant low level of fertility. As a result the zonal soils in this Province are epitomized by Podzols, but in places where the parent materials are rich in calcium, Subarctic Brown Forest, Brown Wooded or Gray Wooded soils have been identified.[390]

The high resin and oil content of coniferous litter and *Betula* bark, together with the normally thick layers of litter and duff, allow devastating fires to spread rapidly through subarctic forest.[2] Areas frequently burned are commonly preempted by *Epilobium angustifolium*, a tall perennial forb with handsome magenta flowers, followed by a shrub stage in which *Rubus idaeus*, *Salix*, and *Vaccinium* provide characteristic dominants. The first forest to rise above these short woody plants may be evergreen (*Pinus banksiana* or *P. resinosa* in the eastern Section; *P. contorta* in the western Section), or deciduous (*Betula papyrifera*, *Populus balsamifera*, *P. tremuloides*, *P. trichocarpa*) (Figs. 23 and 24). Where the duff is too wet to burn completely, deciduous trees succeed better than conifers as invaders. The

Fig. 23. *Picea glauca* and *Abies balsamea* replacing *Pinus resinosa* on coarse, acid glacial drift. Itasca Park, Minnesota.

serotinous cones of *Pinus contorta*, *P. banksiana*, and *Picea mariana* give these species an advantage of an abundant seed supply to invade mineral soil bared by fire. *Picea glauca* usually produces little seed until 35 years old, which places it at a disadvantage as an invader, and if fires follow in short sequence this species can be eliminated from an area.

The arborescent flora of the North American subarctic forest shows close relationships with that of the Eurasian subarctic forest, yet members of the species pairs among the trees are distinct. On the other hand, in the undergrowth there are quite a few species shared by the two hemispheres some being *Linnaea borealis*, *Pyrola secunda*, *P. uniflora*, *Rubus idaeus*, and *Vaccinium myrtillus*.

Nonzonal Vegetation

In late glacial time water draining off wasting ice deposited large sheets of infertile sand in the southern part of the subarctic belt, and such material came to support edaphic climaxes of *Pinus banksiana* or *P. contorta*, with a

Fig. 24. *Picea glauca* replacing *Populus tremuloides*, Yukon Territory.

Fig. 25. *Carex* fen and *Salix* carr, near Snag, Yukon Territory.

sparse ground cover in which fruticose lichens, *Polytrichum,* and a few highly mycotrophic shrubs such as *Arctostaphylos uva-ursi* and *Vaccinium* are predominant.

The glacial debris that was dropped in place was of uneven thickness, and this, combined with the general flatness of the terrain, has resulted in perhaps half the land area supporting lakes, ponds, and mires (Fig. 25). Carex-dominated fens usually encircle any open water that remains, and where the peat surface has been built up above the water table *Sphagnum* forms a ground carpet which at first has only a shrub overstory (*Alnus, Andromeda polifolia, Betula glandulosa, Chamaedaphne calyculata, Kalmia polifolia, Ledum groenlandicum, Rubus chamaemorus, Salix, Vaccinium oxycoccus*) with *Picea mariana* eventually superimposed at climax (Fig. 26). The *Picea* stand is often quite open and the trees much dwarfed (Fig. 27). *Larix laricina* sometimes precedes the entry of *Picea* in this sere, but it tends to be eliminated eventually. *Ledum groenlandicum* and *Sphagnum* are the most characteristic dominants of the shrub and ground layers in bog climaxes (Fig. 28). *Picea/Ledum/Sphagnum* communities are also characteristic of cold north-facing slopes where the permafrost table descends but a short distance in summer. Habitats with only slightly impeded drainage and without peat accumulation occur widely, and these are marked by heavy stands of *Picea mariana* with undergrowth quite similar to that of the *Picea glauca* forests.

Fig. 26. Margin of a young pond near Watson Lake, Yukon Territory. A *Carex* zone next to open water at the extreme right is followed by *Betula glandulosa–Salix* scrub, with *Picea mariana* behind. As yet there has been negligible accumulation of peat.

Fig. 27. *Picea mariana* showing variable and poor vigor in an old bog dominated by *Sphagnum* and ericaceous shrubs, near Millinocket, Maine. The shrub specimens in Fig. 28 were all obtained in the immediate foreground of this photo.

Fig. 28. A collection of ericaceous shrubs all found growing together in the *Picea mariana* bog shown in Fig. 27. Left to right: *Andromeda glaucophylla, Ledum groenlandicum, Vaccinium pennsylvanicum, Kalmia angustifolia, Chamaedaphne calyculata, Gaultheria procumbens,* and *Vaccinium oxycoccus.*

Wetlands better supplied with bases generally remain as marshes, with *Eriophorum* or *Carex* dominating the physiognomy. *Zizania aquatica* is abundant in pond margins.

In the Peace River drainage of northern Alberta the upland forests were frequently interrupted by lush meadows dotted with patches of *Symphovicarpos-Rosa* scrub, and groves of *Populus tremuloides* or *Salix*.[298] The principal graminoids of these parks were *Agropyron trachycaulon*, *Carex*, and *Stipa spartea curtiseta*, mingled with a variety of forbs none of which gained dominance.

Most of these parks represent the beds of former shallow ponds, and their heavy black soil provides choice agricultural land in a region where podzolization of coarser soils limits this pursuit.

In the Yukon drainage xerophytic parks are conspicuous on slopes that face steeply to the south. *Calamagrostis purpurascens*, *Festuca altaica*, and *Artemisia frigida* are common dominants. Again a belt of *Populus tremuloides* separates the parks from conifer forest.

Characteristic vertebrates of this Province include moose, woodland caribou, white-tailed deer, black-tailed deer, porcupine, beaver, varying hare, red squirrel, chipmunk, muskrat, black bear, wolverine, lynx, red fox, wolf, fisher, otter, mink, marten, weasel, grouse of several kinds, pine grosbeak, cross-bill, junco, Canada jay, and hawk-owl. The red squirrel bids fair to be the most influential vertebrate owing to its prodigious consumption of tree seeds (Fig. 29), especially since it prefers certain species over others.

Fig. 29. Middens about a squirrel cache in *Picea engelmannii–Abies lasiocarpa* forest, Wyoming. The total range of this squirrel, *Sciurus hudsonicus*, in North America roughly parallels that of the Subarctic–Subalpine Forest Region.

Land Use

Although *Picea glauca* is a good timber species, a major use of this tree is for paper pulp. In the northern half of the Region, including the broad ecotonal area, the trees are too small and too inaccessible to be valuable for their wood. In those latitudes extensive herds of caribou depend upon epiphytic lichens and shrub browse for their winter sustenance. Meat of the reindeer (European relative of caribou) is an important economic resource in Eurasia. Trapping, hunting, fishing, and other recreational pursuits are additional important uses of the natural ecosystems.

Agriculture is limited by the short summers and infertile soils, yet there is a significant production of white potato, cabbage, turnips, rutabagas, carrots, peas, flax, rape, oats, hay, and pasturage. Nitrogen and phosphorus fertilizers are needed regularly, and calcium is often deficient as well. Some beef cattle are produced in North America, but milk cattle are emphasized in northwestern Europe.

The Ecotone with Tundra[209]

In different places around the Northern Hemisphere arctic timberline may be formed by evergreen trees (*Picea* or *Pinus*) or by deciduous trees (*Larix* or *Populus*). From Labrador to the Mackenzie Delta *Picea mariana* is the most ubiquitous tree, although *Abies balsamea, Larix laricina,* or *Picea glauca* occur locally. From the Mackenzie Delta to western Alaska *Picea glauca* is the prevailing timberline tree, with *Populus balsamifera* occasionally occupying this position. At the base of the Alaska Peninsula *Picea sitchensis* forms timberline.

The ecotone between subarctic forest and tundra is a very broad belt in which the two vegetation types often alternate, with the proportion of forest to nonforest vegetation changing progressively across its breadth. Starting from a point well within developed forest and progressing northward one passes in turn: (1) the limit beyond which natural regeneration of forest is so slow that logging is inadvisable, (2) the limit of continuous forest, (3) the farthest individuals that might be classed as trees (3 m tall?), and (4) the last individuals of species which have the potential of trees, but which have been reduced to shrub stature. Low tree density in the last few of these belts reflects rarity of seed production and of germination success; dwarfness reflects the impact of heat deficiency and winter killing of incompletely hardened tissues.

Since local conditions of climate or soil have appreciable influence by aggrevating or ameliorating the increasing environmental stress for woody vegetation, these physiognomic components of the broad ecotone are sinuous. Trees extend farthest north along the sides of valleys of the streams that flow into the Arctic Ocean, and in the Mackenzie Valley forest actually reaches the coast of that Ocean, East of Hudson's Bay the development of continuous forest has been prevented in large areas by a lack of soil, for trees are limited to groves growing where lenses of gravelly soil have accumulated in depressions in ice-scoured bedrock.[276] West of the Bay such soil limitation is of minor importance, and there the trees are often scattered individually as in a savanna, with a ground cover of gray lichens (*Cladonia, Stereocaulon,* etc.) up to 15 cm thick, dotted with shrubs (*Betula glandulosa, Empetrum, Ledum groenlandicum, Vaccinium*).

This broad mosaic ecotone is often mapped separately.

The Ecotone with Steppe

From northern Minnesota to central Saskatchewan and west to the Rockies, the *Picea glauca* Province abuts the Steppe Region. Adjacent to the forest here is a belt of *Populus tremuloides* groveland which at this margin consists of essentially continuous *Populus* forest.

The Ecotone with Temperate Mesophytic Forest

From Minnesota to Maine taiga abuts upon temperate mesophytic forest. In this ecotone conifer stands become confined to coarse-textured soils low in bases where their metabolism promotes podzolization which they can tolerate. Deciduous forest, in contrast, is favored by fine-textured soils where by recycling bases they can maintain the high level of fertility which they require. In addition to this mosaic of distinct types, there are associations in which dominants of the two major vegetation types are on an equal footing, especially communities in which *Abies balsamea* and *Acer saccharum* share climax status.

At the base of the Kenai Peninsula in Alaska, and in the Skeena River Valley of British Columbia, the *Picea glauca* Province makes very limited contact with the *Tsuga heterophylla* Province.

Ecotones with the Subalpine Provinces

At the northern end of the Appalachian Mountains, the ecotone between the *Picea glauca* and *P. rubens* Provinces is marked by forests in which dominants of the two grow together. For example, *P. rubens* and *Abies balsamea* form one of the hybrid types.[111]

From approximately central Alberta and British Columbia northward, *Picea glauca* forest not only dominates the basal plains, but extends up the mountain sides to alpine timberline. South of here the *Picea engelmannii* Province occupies the subalpine belt of the Rockies, with *Picea glauca* associations becoming restricted to lower elevations on the same slopes. There, hemmed in by *Picea engelmannii* forests above and by xerophytic forest or steppe below on the basal plains, *Picea glauca* forests dwindle southward, disappearing at approximately the international border. Throughout British Columbia and Alberta as well as in the adjacent United

States, the contact of *Picea glauca* and *P. engelmannii* is marked by extensive hybridization of the two species.[107]

Near the Pacific coast the tundra-covered summits of coastal mountain ranges separate forest of *Picea glauca* in continental climates on the inland side, from subalpine *Tsuga mertensiana* forests in oceanic climates of the slope facing the Pacific Ocean.

The *Picea rubens* Province[87,235,310]

Forests closely related to those of the subarctic *Picea glauca* belt occur in the Appalachian Mountains from Maine to Tennessee and North Carolina, except for gaps, especially in Pennsylvania, where the mountains are too low (Fig. 30). This is the smallest Province of the Subarctic–Subalpine Forest Region, yet two sections can be recognized.

High on mountain slopes in the northern Section, *Abies balsamea* and *Picea mariana* are the major climax dominants with lesser amounts of *Betula papyrifera* in the stands (Fig. 31).[367] At lower altitudes *Abies* shares dominance with *Picea rubens* (Fig. 32). The undergrowth, in general, includes an

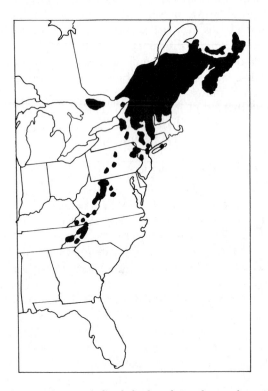

Fig. 30.   Range of *Picea rubens*, including hybrids with *P. galuca* in the northern extremity.

Fig. 31. *Picea mariana–Abies balsamea* forest on Mt. Katahdin, Maine. In this forest, where the taxonomic affinities are about equally divided between the *Picea glauca* Province and the *Picea rubens* Province, *P. mariana* demonstrates its capacity to become an important upland tree, especially in mountains.

Fig. 32. *Picea rubens–Abies balsamea* forest on Mt. Desert Island, Maine. Bryophytes form an almost complete ground cover in this association, which has taxonomic affinities about equally divided between the *Picea glauca* Province and the *Picea rubens* Province.

abundance of ferns and many seed plants showing close affinities with the subarctic forest: *Clintonia borealis, Cornus canadensis, Linnaea borealis, Mianthemum canadense, Oxalis montana, Sorbus americana,* and *Viburnum edule,* along with the more abundant *Nemopanthus mucronata, Vaccinium angustifolium, V. erythrocarpum,* and *Viburnum alnifolium.* Natural hybrids between *Picea rubens* and *P. glauca* occur in the northern end of the Province.

Epilobium angustifolium and *Pteridium aquilinum* are well represented on burned land in this northern Section, with these followed by shrubs such as *Ribes* and *Rubus,* then seral forests composed variously of *Acer rubrum, Betula papyrifera, B. populifolia, Pinus resinosa, Populus grandidentata, P. tremuloides,* and *Prunus virginiana.* Repeated burning tends to favor ericaceous shrubs, especially *Vaccinium* and *Gaylussacia baccata.*

Here in the north *Picea mariana* plays another role, occurring in bogs where it may be associated with *Larix* and bog shrubs that are common in the subarctic belt.

In the southern Section the arborescent flora is much poorer. Low in the subalpine belt *Picea rubens* tends to dominate in pure stands, with *Abies fraseri* occurring alone at high elevations, and in mixed communities between (Fig. 33). Although *Betula alleghaniensis* is basically a seral tree, it persists in low densities in climax forests (Fig. 34).

Rainfall up to about 2300 mm (Fig. 11F), combined with occasional summer fog, creates the high humidity that favors essentially complete coverage

Fig. 33. *Picea rubens–Abies fraseri* stand in Great Smoky Mountains National Park, Tennessee.

Fig. 34.   Forest dominated by *Picea rubens, Abies fraseri,* and *Betula alleghaniensis* (winter condition) on a high ridge in the southern Appalachian Mountains.

of epigeous and corticolous bryophytic unions at least high in the belt. The herbs and shrubs still show close relations with the subarctic forests by including such plants as *Acer spicatum, Clintonia borealis, Maianthemum canadense, Oxalis montana, Sambucus pubens, Sorbus americana, Streptopus roseus, Vaccinium erythrocarpum, Veratrum viride,* and *Viburnum alnifolium.* Following fires *Betula* and *Viburnum* sprout from the root base, and are joined by *Rubus canadensis* and *Prunus pennsylvanica.*

Podzol and Brown Podzolic Soils still prevail in these subalpine forests.

Here in the southern Appalachians *Picea-Abies* forests occur only in those districts where there are peaks rising above about 1740 m. Apparently these trees were eliminated from other districts where there are no peaks high enough to have served as refugia when ecotones were elevated during Hypsithermal time. If this hypothesis is correct, the lower forest ecotone rose approximately 400 m, and this would imply a Hypsithermal rise in temperature of 1.49°C if the present lapse rate of 4.03°C/100 m prevailed at that time.[358]

A curious type of vegetation that breaks the continuity of coniferous vegetation in the southern Appalachians consists of islandlike stands of deciduous trees, *Fagus grandifolia* or sometimes *Betula alleghaniensis,* or *Aesculus octandra,* which occupy saddles in ridges otherwise occupied by conifers

Fig. 35.   A saddlelike stand of *Fagus grandifolia* forest, in leafless winter condition, extending across a low point on a ridge otherwise dominated by *Picea rubens–Abies fraseri* forest in the southern Appalachian Mountains.

(Fig. 35). These probably owe their existence to microclimates dependent on air movement across the saddles.[348]

Another special feature of the vegetation here in the south occurs on upper slopes and ridges fully exposed to wind. These are often treeless, supporting dense thickets of evergreen ericaceous shrubs among which *Kalmia latifolia, Rhododendron carolinianum, R. catawbiense, R. punctatum,*

Fig. 36.   Ericaceous shrubs dominating sharp ridges directed southwest in the subalpine forest of the southern Appalachians of Tennessee.

and *Vaccinium corymbosum* are conspicuous dominants (Fig. 36).[60] Since the highest peaks are usually forested, and these shrubby parks commonly extend down into the edge of deciduous forest below the subalpine belt, this scrub has no ecologic relation to the ericaceous scrub that is so common in the lower margin of alpine tundra. These habitats are often burned, but the wet peat is seldom charred deeply between rains, and the shrubs sprout promptly. Locally some botanists have hypothesized replacement of such scrub by forest, but no firm evidence based on populations structure has been presented. Most of these shrub communities seem best regarded as topographic climaxes.

Ecotones

From Maine to New York the *Picea rubens* Province enmeshes small islands of alpine tundra atop a few of the highest mountains. Here as in other mountains of North America this ecotone between forest and alpine tundra is very narrow in comparison with the situation at arctic timberline.

On lowlands across the northern end of the *Picea rubens* Province it intergrades with the *Picea glauca* Province by means of associations with dominants drawn from both.

Southward *Picea rubens* forests form ecotones with the *Acer saccharum* Province lower on the mountain sides, these involving principally associations in which *Acer*, *Fagus*, *Tsuga canadensis*, and *Betula alleghaniensis* are the major climax dominants.

Well to the north of the Appalachian Mountains there are low mountains which rise above tree limits and have well-developed alpine vegetation.[197] *Picea glauca*, *P. mariana*, and *Larix laricina* of the subarctic forest extend up to timberline here, without an intervening more specialized subalpine belt.

The *Picea engelmannii* Province

The *Picea engelmannii* Province occurs as a belt just below alpine tundra in the Rocky Mountains, from central Alberta and British Columbia to Arizona (Fig. 37). The Rockies are close enough to the Pacific Ocean that air masses following the path of the Westerlies crossing these mountains in the vicinity of the Canada–United States border strongly affect climates as far inland as the continental divide. Here in the north where oceanic influences are dragged inland they favor a special floristic element which is superimposed on the more widespread montane flora of the Rockies, and which serves to differentiate a northern from a central Section lacking this element. At approximately the southern boundary of Colorado another floristic break sets off a southern Section.

In the central Section *Abies lasiocarpa* is the characteristic dominant of nearly all the climatic climaxes. *Picea engelmannii*, which is only slightly less

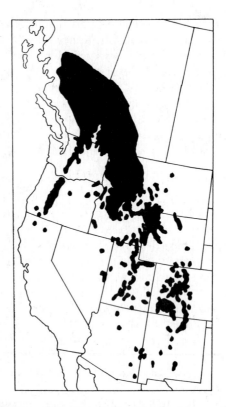

Fig. 37. Range of *Picea engelmannii,* including the extensive area of hybridization with *P. glauca* in British Columbia, Alberta, and western Montana. Since this tree extends somewhat beyond the limits of the *Picea englemannii* Province on all margins, the map locates the Province but exaggerates its coverage.

tolerant of competition but is a much taller and longer-lived tree, usually plays a late-seral role. But in certain habitats it seems capable of retaining a place in the climax indefinitely. Rarely it is the sole climax dominant. In the undergrowth the dwarf shrub *Vaccinium scoparium* is widespread and very abundant, with the sparse forb component usually including *Arnica cordifolia.* In places a richer undergrowth type includes shrubs such as *Ribes montigenum, Lonicera utahensis,* and a variety of herbs. The soils range from Podzol to Gray–Brown Podzolic to Brown Podzolic.

Following fire the most aggressive seral tree is *Pinus contorta* (Figs. 38 and 39), but *Populus tremuloides* plays a similar role on relatively moist sites, and low in the subalpine belt *Pseudotsuga menziesii* takes advantage of deforested areas temporarily.

At upper timberline (Figs. 40 and 41) are groves of wind-dwarfed and deformed trees composed of various combinations of *Abies lasiocarpa* (Fig.

Fig. 38.   Where old trees of *Pinus contorta* are represented in a forest that burns, highly successful reproduction from seeds held in serotinous cones frequently results in "dog hair" stands of the tree.

Fig. 39.   *Abies lasiocarpa* replacing *Pinus contorta*, with *Vaccinium scoparium* dominating the undergrowth. Targhee National Forest, Wyoming.

Fig. 40. Upper timberline in Rocky Moutain National Park. The frosty valley bottom supports *Salix* scrub. As usual, *Abies lasiocarpa* and *Picea engelmannii* show no stunting influence of the worsening climatic gradient up to the margin of a relatively narrow belt of krummholz.

Fig. 41. The distribution of trees at upper timberline may be strongly influenced by differences in soil or topography as it affects the depth of snow. Note restriction of trees to crests of ridges at lower left. Beartooth Mountains, Montana.

Fig. 42. *Abies lasiocarpa* at upper timberline showing strong difference between the basal mat derived from layered branches protected by snow in winter, and the part that projects above the snow. In this ecotone the tree clones are dispersed over a continuous phase of alpine meadow vegetation. Medicine Bow Mountains, Wyoming.

42), *Pinus flexilis, P. aristata,* and *P. longaeva,* the last attaining an age exceeding 7000 years in this marginal arboreal environment.

The southern Section differs from the central Section with the absence of *Pinus contorta* and *Vaccinium scoparium,* and in the substitution of *Abies lasiocarpa arizonica* for *A. lasiocarpa lasiocarpa.*

The northern Section is basically similar to the central Section except for the added oceanic floristic element. Here *Tsuga mertensiana* is a strong competitor that plays a climax role locally, since it can overwhelm *Abies lasiocarpa* as well as *Picea.* In the undergrowth *Menziesia ferruginea, Rhododendron albiflorum, Sorbus sitchensis, Vaccinium membranaceum,* and *Xerophyllum tenax* become important dominants that reduce *Vaccinium scoparium* to a relatively minor role, especially on slopes facing into the Westerlies. Even trees that play a seral role are enriched by *Larix occidentalis* and *Pinus monticola.* At upper timberline *Abies lasiocarpa* shares the ecotone locally with only *Pinus albicaulis* or *Larix lyallii.*

The flora of the *Picea engelmannii* Province as a whole shows close relationships with the *P. glauca* Province to the north. Not only do *Abies lasiocarpa, Picea engelmannii,* and *Pinus contorta* hybridize where they come into contact with their homologs *Abies balsamea, Picea glauca,* and *Pinus banksiana,* but a significant number of undergrowth plants are shared, including *Clintonia, Cornus canadensis, Juniperus communis, Linnaea borealis, Monotropa hypopitys, Pyrola secunda, Shepherdia canadensis,* and *Sorbus.* However, the absence of paper-barked *Betula* in the upland forests of the Rockies is notable.

In the central Section, at least, *Pinus contorta* appears to have a climax status in habitats where a generally impoverished or xerophytic undergrowth suggests locally infertile soils. Elsewhere, on unusually moist and fertile soils *Populus tremuloides* becomes a self-perpetuating tree, with a lush undergrowth of tall forbs where cattle have not destroyed it.

On wind-swept ridges *Pinus flexilis* or *P. albicaulis* may form pure forests which, although not tall of stature, are not as dwarfed and misshapen as in the timberline groves. In the latter, these light-demanding pines achieve climax status along with *Abies lasiocarpa* only because the arborescent stratum cannot close over.

Parks of many kinds interrupt forest.[126] Seasonally wet parks have heavy stands of special grasses and sedges, often with a shrub component of *Salix, Lonicera involucrata,* or *Potentilla fruticosa* as well. Those on well-drained valley fill are grasslands, if such be the character of the adjacent basal plain, or shrub-steppe with *Artemisia* where taxa in this genus are well represented on the adjacent basal plain.

Sphagnum bogs supporting ericads such as *Kalmia polifolia, Ledum glandulosum Vaccinium caespitosum, V. occidentale,* etc., characterize wet sites where fertility is low. Deep snow accumulations or frostiness due to cold air drainage usually prevents trees from growing in the hollows as upper timberline is approached (Fig. 40).

Some representative vertebrates of the Rocky Mountain Subalpine Forest belt include wapiti, deer, mountain goat, bighorn sheep, beaver, porcupine, grizzly bear, black bear, wolf, coyote, puma, lynx, bobcat, otter, marten, weasel, wolverine, grouse of several types, Canada jay, and Clarks crow.

Although *Picea* lumber is of good quality and is regularly harvested, the low value of *Abies* and general inaccessibility of much of the forest in this Province makes it more valuable for holding snow, providing water for irrigation on the dry basal plains. Recreation values, especially skiing, are very high.

At its northern terminus the *Picea engelmannii* Province abuts the transcontinental *P. glauca* Province. As this ecotone is approached, *P. mariana* becomes an additional member of the upland climax forest.

At its upper limits the Province forms an ecotone with alpine tundra, if the mountain has sufficient altitude, and below it usually forms an ecotone with the *Pseudotsuga menziesii* Province. In the northern Section the *Tsuga heterophylla* Province is inserted between the *Picea* and *Pseudotsuga* Provinces.

In a few places the normal zonal sequence of forest belts on the mountain sides is truncated from below so that steppe extends up to the *Picea* belt.

The *Tsuga mertensiana* Province

The *Tsuga mertensiana* belt occupies a belt just below upper timberline in the Sierra Nevada, Cascade and other ranges that continue northward along the Pacific coast, descending gradually to near sea level at approximately Cook Inlet, Alaska (Fig. 43).

This Province is altitudinally homologous with the *Picea engelmannii* and *P. rubens* Provinces since it occurs just below alpine tundra, and like them has short cool summers alternating with long snowy winters. It differs from those Provinces, as well as from the *Picea glauca* Province, in that the precipitation is restricted mainly to winter, and may be lower as well (compare Fig. 11H with 11E–G). Since the climatic requirement of most plants of the subarctic forest prevented them from spreading into the mountains near the Pacific coast, a special subalpine flora evolved from the temperate

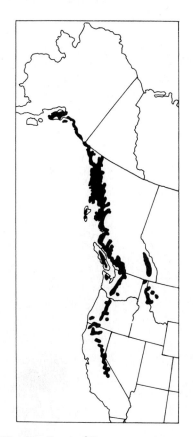

Fig. 43.   Range of *Tsuga mertensiana*.

mesophytic forest of the lowlands, as these mountains uplifted in Pliocene time.

In a northern Section that extends the length of Alaska from the Kenai Peninsula southward along the Pacific slope, this Province is represented by climax forests in which *Tsuga mertensiana* is accompanied by only minor amounts of *Chamaecyparis nootkatensis* at most. Conspicuous undergrowth plants include *Cladothamnus pyrolaeflorus, Menziesia ferruginea, Rubus pedatus, Sorbus sitchensis, Vaccinium ovalifolium,* and *V. uliginosum. Picea sitchensis,* the principal seral tree here, extends to the alpine timberline in places where it achieves climax status as stunted groves of trees (Fig. 44).

A central Section of the Province extends from approximately the southern tip of Alaska down the western margin of British Columbia, then follows the Cascades across Washington and Oregon.[145] The arborescent flora here is relatively rich, the principal climax trees of the different habitat types being *Abies amabilis, A. lasiocarpa, Chamaecyparis nootkatensis,* and *Tsuga mertensiana* (Fig. 45). Just below upper timberline *Tsuga* dominates most of the climaxes on seaward slopes, with *Abies lasiocarpa* locally dominant on landward slopes and *A. amabilis* taking over this role slightly lower in the subalpine belt. Representative undergrowth plants here include those mentioned for Alaska plus *Pachistima myrsinites, Rhododendron macrophyllum, Vaccinium membranaceum,* and *Xerophyllum tenax.* These communities are associated with Podzol, Grey Podzolic, and Brown Podzolic Soils.

Fig. 44. Ecotone between subalpine forest and alpine tundra, above Haines, Alaska. Scattered *Picea sitchensis* in foreground, with the shrub belt in adjacent tundra dominated by *Alnus.*

Fig. 45. *Abies amabilis–Tsuga mertensiana* forest on a high ridge in the Cascade Mountains of Washington.

The seral trees are *Abies lasiocarpa* (in certain habitat types), *A. procera, Larix occidentalis, Picea engelmannii, Pinus contorta, P. albicaulis* (near upper timberline), *P. monticola,* and *Pseudotsuga menziesii.* In the timberline ecotone *Larix lyallii, Pinus albicaulis,* and *P. balfouriana* maintain places as climax species, along with *Abies lasiocarpa* and *Tsuga,* although only two or three of these are usually present in any one district (Fig. 46).

It is notable that in this central Section of the Province, which is obviously differentiated by the strong influence of the Westerlies moving eastward in this latitudinal range, there is the maximal interchange between plants of

Fig. 46. *Abies lasiocarpa* and *Tsuga mertensiana* confined to ridge summits at upper timberline on Mount Rainier, Washington.

oceanic and continental climates. For example, only here do *Larix lyallii, L. occidentalis, Pinus albicaulis,* and *Tsuga mertensiana* also occur in the Rockies, and only here do the continental species *Abies lasiocarpa* and *Picea engelmannii* penetrate westward strongly into oceanic climates.

As in subalpine forests of the Rockies, a wide variety of edaphically determined parks disrupt the continuity of forest (Fig. 47).

The southern Section, from southern Oregon to southern California along the Cascade–Sierra axis, has a tree flora that is considerably less rich, with *Abies magnifica* dominant in the lower part of the forest (Fig. 48), and *Tsuga mertensiana* dominant above. Herbaceous and shrubby vegetation in these forests is rather sparse. It includes only a few plants shared with the *Picea glauca* Province, such as *Cornus canadensis, Linnaea borealis,* and *Pyrola,* growing with *Arctostaphylos patula, Castanopsis, Ceanothus, Chrysopsis, Holodiscus microphyllus, Monardella, Prunus, Ribes,* and *Rubus.*[309] *Arctostaphylos, Castanopsis,* and other evergreen shrubs proliferate rapidly after fires.

Seral trees include *Pinus attenuata, P. contorta, P. monticola, P. albicaulis* high in the belt, and *Abies concolor* low in the belt.

Clonal stands of *Populus tremuloides* are characteristic of moist pockets of fertile soil, and *P. fremontii* may be found on streambanks throughout the Province.

At upper timberline *Pinus albicaulis, P. balfouriana, P. flexilis,* and *P. longaeva* commonly join *Tsuga mertensiana* in the krummholz.

Fig. 47. Lush meadow in subalpine forest belt, Mount Rainier, Washington. *Erigeron, Ligusticum, Lupinus,* and *Polygonum* share dominance with grasses and sedges.

Fig. 48. *Abies magnifica* forest. California.

Wherever mountains rise to sufficient height, alpine tundra occurs immediately above the *Tsuga mertensiana* Province. At its lower altitudinal limits the latter forms an ecotone with temperate mesophytic forest—with the *Tsuga heterophylla* Province from Alaska to central Oregon, and with the *Pseudotsuga menziesii* Province from there southward.

The *Pinus hartwegii* Province

Savannalike stands of *Pinus hartwegii* characterize a forest belt just below the alpine zone of high mountains in Mexico. The limited amount of information available indicates that snow falls on this subalpine belt each winter, but each snowfall melts shortly.

The herbaceous ground cover of these open forests is composed of species in genera which are common in temperate latitudes, such as *Agrostis, Festuca, Muhlenbergia, Ranunculus,* and *Tauschia.* Among the scattered shrubs are *Juniperus tetragona* and *Pernettya ciliata.* In the lower part of the subalpine belt *Abies religiosa* occurs in pure stands or in mixtures with *Pinus ayacahuite* or *P. montezumae,* and here too the undergrowth is strongly related to the flora of temperate North America: *Alnus, Dryopteris, Epilobium, Erigeron, Gentiana, Monotropa, Montia, Salix, Senecio, Trisetum,* and *Verbena.*[349]

At the lower margin of the subalpine belt it forms an ecotone with forests dominated by *Quercus,* with various species of *Pinus* in seral or edaphic climax roles, or on fog-shrouded slopes forms an ecotone with richer forests having tropical genera well represented, especially in the undergrowth.

At upper timberline *Pinus hartwegii* is dwarfed but not misshapen, and the highest individuals are located on ridges. In this ecotone shorter conifers may be locally dominant, these including *Juniperus mexicana, J. monticola, J. tetragona, Pinus culminicola,* or *P. flexilis.*

Pinus hartwegii has good form and quality, and so is exploited for lumber.

THE TEMPERATE MESOPHYTIC FOREST REGION

The Temperate Mesophytic Forest Region includes those areas in temperate latitudes where zonal soils remain moist to near the surface year-round, and where frost is a regular feature of winter, yet there is a relatively long frost-free summer. The trees are mostly deciduous angiosperms where the summers are hot (Fig. 11I,J), and mostly evergreen conifers where they are cool (Figs. 11K,L and 12A,B). Approximately 50% of the temperate mesophytic forest flora is hemicryptophytes, with phanerophytes, geophytes; and therophytes each contributing about 15% of the total species. The coniferous and most of the angiosperm trees are wind pollinated, and come into flower at the start of the warm season. In both physiognomic types of forest there is a nearly closed layer of tall trees, with a diffuse lower layer of short sciophytic trees. Where angiosperm trees dominate, only about 25–38% of the woody plants have entire-margined leaves or leaflets.

Origin and Distribution

Genera of trees characteristic of the modern temperate mesophytic forest are well represented in Cretaceous sediments, but nowhere do they appear to have been dominant.[435] At the time of maximum Cenozoic warmth late in the Eocene Epoch they were still widespread in mixture with tropical genera, but by this time a belt of purely temperate flora had segregated out across the northern border of North America where it was free of tropical associates. At this time the temperate flora became circumpolar by way of Beringia and North Atlantic islands.

The temperate flora is thought to have been derived from genera originally represented only in the margin of a broad belt of tropical forest. This is suggested by three types of evidence. (1) The most primitive species of *Acer, Betula, Celtis, Populus, Prunus,* and *Quercus,* which are primarily characteristic of northern temperate forests, are largely evergreen plants of the tropics or subtropics.[17] (2) Many genera represented by only herbs in extra-tropical areas have woody relatives in the tropics, and herbaceous plants have evolved from woody ancestors. (3) The temperate mesophytic forests of both northern and southern hemispheres show many floristic relationships with the tropical forest flora that separates them, but they show no direct relationships to each other.

With the start of Cenozoic cooling tropical elements receded southward more rapidly than did temperate elements. Even as late as Miocene time a coastal strip of temperate forest extended as far north as Beringia, and during that Epoch it was mainly just the deciduous element that receded from that area. Modern temperate forest floras of North America and Eurasia have many closely related vicariads that have essentially identical ecologies, and therefore must have evolved from a common circumpolar ancestor through the accumulation of mostly nonessential characters after contact across the oceans was lost.[234]

Conifers have been components of the temperate mesophytic forest flora throughout Cenozoic time, but they have been well represented only in the cooler parts at higher latitudes or altitudes. For example, in southwestern Alaska, fossils representing temperate forests of Miocene age include *Fagus*, *Liquidambar*, *Quercus*, and *Tilia* with *Abies*, *Chamaecyparis*, *Picea*, *Pinus*, *Thuja*, and *Tsuga*. Down near the sea the angiosperm elements appear to have been dominant over the conifers, but on mountains a short distance away this relationship was reversed.[437]

The spread of aridity as the temperate mesophytic forest moved southward resulted in its segregation into three areas. One of these came to occupy roughly the southeastern quarter of North America, separated by a broad midcontinental arid tract from another segment in the Cordillera to the west. The third segregate consisted of those elements that had migrated so far south down the Cordillera by mid-Miocene time that intensifying aridity along the Mexico–United States border cut them off as a tropical segment of the Cordilleran flora.[160]

The enforced southward migration during Cenozoic cooling involved so much time, distance, and consequently, diversity of climates, that this stress, coupled with genetic discontinuities, resulted in a high degree of differentiation of a formerly somewhat homogeneous and rich holarctic flora. Then the severe climatic oscillations of Pleistocene time must have been effective in reducing the number of ecotypes which had permitted species and genera of earlier epochs to be represented over wide area.

Parallel events in Eurasia during the Cenozoic Era split the temperate mesophytic forest on that continent too, with one unit left centered on the Elbe Valley, another centered on Korea, and with montane fragments stranded on the Himalayas and other southeast Asian mountains. As in North America, these remnants are separated by vast expanses of xerophytic vegetation.

In the Southern Hemisphere the most homologous vegetation is the mainly evergreen dicot forest characterized by *Nothofagus* that occurs now in southern Chile and New Zealand. In Cretaceous or Eocene time this forest also occurred on Antarctica and nearby islands, then later lost connection across the South Pacific Ocean either in consequence of continental drift or Cenozoic cooling that enforced contraction toward the equator.

During Wisconsin glaciation certain temperate forest trees of eastern North America persisted within 60 km of the ice front in Illinois,[167] with similar situations in Pennsylvania[277] and New Jersey. This clearly indicates a very steep temperature gradient that probably resulted from the Westerlies which kept the cold air that flowed off the ice swept eastward over the Atlantic Ocean rather than spreading far southward. The fact that all four known continental glaciers extended only as far south as this storm track strongly suggests its significance as a major controlling influence. However, temperatures to the south of the ice were cooled sufficiently that the more sensitive temperate zone species were crowded to the southern tier of states, with subarctic conifers infiltrating that far south. Relictual northern elements including *Taxus floridiana* in the northern Florida,[243] indicate clearly that the ranges of some of the temperate forest elements were strongly displaced, even though some proved remarkably preadapted to tolerate chilly environments just south of the ice margin. That remarkable mingling of trees now well segregated latitudinally was probably a consequence of increased oceanicity of climates during the Wisconsin glaciation.[253]

In the middle of the continent Wisconsin ice expanded so rapidly that it probably advanced directly onto steppe in the north, but with precipitation becoming more effective the remaining steppe was slowly invaded by trees. Subarctic species spread southward almost to the Gulf of Mexico, montane forest trees expanded eastward at least to the Dakotas, and deciduous trees spread westward to the foot of the Rockies. A relic population of *Acer saccharum* that has persisted in western Oklahoma (Fig. 49), and a swarm of relics left stranded in the Black Hills of South Dakota, suggest a mass advance of temperate mesophytic forest westward. The Black Hills relics include *Celastrus scandens, Celtis occidentalis, Corylus americana, Fraxinus pennsylvanica, Monarda fistulosa, Ostrya virginiana, Parthenocissus quinquefolia, Quercus macrocarpa, Smilax herbacea, Ulmus americana,* and *Vitis vulpina*—all clearly characteristic of deciduous forests east of the steppe.

Once the Wisconsin icecap began continuous recession, about 11,000 BP, forest expansion northward over the glaciated area was rapid, since *Pinus banksiana, Picea glauca,* and *Betula papyrifera* are especially favored by bare mineral soil, and competition among the widely spaced pioneers was negligible. On the other hand, among the temperate forest species northward migration was undoubtedly slower since most of the area involved had continued to support forest of mixed affinities and competition had to be overcome. It is generally believed that any range adjustments of temperate forest species that may be in progress at present are more likely a consequence of temporary Hypsithermal retraction or of the more recent warming of climates.

The above remarks refer mainly to that segment of the temperate mesophytic forest located in southeastern North America. In the Cordilleran

Fig. 49. Range of *Acer saccharum*. Isolated localities indicate range withdrawal from the west and from the south in post-Wisconsin time.

segment there is evidence of temporary Wisconsin displacement southward along the Pacific coast, and to lower altitudes inland.[108]

Environment

The climates of the Temperate Mesophytic Forest Region are distinguished from those of the Subarctic–Subalpine Forest Region primarily by a longer frost-free season with more heat units received in summer (compare Figs. 11E,F,G,H with 11I,J,K,L, and Fig. 12A,B). Winters can be severely cold, as in Wisconsin or may have only a few days of freezing weather as in central Florida.

In southeastern North America the Westerlies provide precipitation in winter, and tropical air pushing northward off the Caribbean brings moisture in summer. In consequence, zonal soils remain moist to near the surface the year-round.

Along the Pacific coast temperate mesophytic forest is restricted to latitudes where there is much rain and cloud in winter as the Westerlies drag marine air onto land. A northward retreat of the Westerlies in summer leaves much of the central and southern part of the coast without rain at this season, but coolness, and especially the almost daily cloudiness if not fog toward the south, minimize evapotranspiration so that the winter store of soil moisture carries over until the autumn rains begin. Owing to the oceanic climate, many days in winter are above freezing, and evergreen trees can take advantage of such periods to compensate for low temperatures in summer.

Much-attenuated remnants of this western segment of the temperate mesophytic forest have also persisted inland in the Cordillera as narrow belts on mountain slopes, hemmed in by aridity below and deficient heat above.

That segment of the temperate mesophytic forest which reached Mexico and Central America before aridity severed its connection to the northern segment, occurs on midmontane slopes, with heat deficiency setting the upper limits and either drought or insufficient cold stimulation limiting below.

The term "temperate rainforest" has at times been applied to parts of the *Nothofagus* forest of the Southern Hemisphere, and to parts of the temperate mesophytic forest along the Pacific coast of North America. This segregation would seem to have little merit, for the North American "rainforest" tract is in reality very little different in either flora or vegetation from adjacent areas. Higher productivity and an abundance of epiphytes seem to be the chief distinguishing attributes.

Wherever evergreen trees dominate temperate mesophytic forest there is an appreciable accumulation of litter and duff since the firm-textured foliage requires more than 1 year to decay. However, where deciduous trees dominate the leaves cast at the end of summer are almost completely decomposed by the time the next year's crop of leaves is cast. The sudden leaf cast of dicot trees in autumn so completely blankets the soil that epigeous cryptogams are smothered, except on the summits of boulders or on tree bases where the leaves cannot lodge. Owing to the more protracted cast and the slenderness of conifer needles, epigeous bryophytes can grow upward rapidly enough to prevent their smothering, so that evergreen forests usually have an abundance of ground-dwelling bryophytes, and often lichens as well.

The *Acer saccharum* Province

The two Provinces of the Temperate Mesophytic Forest Region that occupy most of southeastern North America have much in common in their physiognomy.[49] Most of the trees are winter-deciduous dicots. Their leaves are membranous, and either toothed or lobed if not both (Fig. 50). Specialized bud scales are well differentiated, and the trunks have rough bark.

Trees of the dominant stratum average about 30 m in height, with the upper surfaces of their canopies remarkably concordant despite the considerable number of species growing intermingled. The highly discontinuous understory of low trees includes one or more of the widely distributed

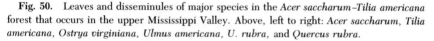

Fig. 50.   Leaves and disseminules of major species in the *Acer saccharum–Tilia americana* forest that occurs in the upper Mississippi Valley. Above, left to right: *Acer saccharum, Tilia americana, Ostrya virginiana, Ulmus americana, U. rubra,* and *Quercus rubra.*

Carpinus caroliniana, Cornus florida, and *Ostrya virginiana,* to which other species of similar ecology are added locally. Beneath these a rather rich shrub component is comprised of such taxa as *Euonymous, Hamamelis virginiana, Lindera benzoin, Parthenocissus quinquefolia, Rhus toxicodendron, Taxus canadensis,* and *Viburnum. Celastrus scandens* and *Vitis* are common lianas, especially on river terraces where they may be joined by liana forms of *Parthenocissus quinquefolia* and *Rhus toxicodendron.* Some representative herbs are *Actaea rubra, Anemonella thallictroides, Arisaema, Claytonia virginiana, Dicentra, Erythronium, Hepatica triloba, Mitella diphylla, Phryma leptostachya, Sanguinaria canadensis, Smilacina, Solidago, Trillium,* and *Viola.* Some of these herbs are heliophytes that have a burst of flowering and vegetative activity while the tree foliage is expanding in spring, the plants then going into aestivation. Other herbs, obligate sciophytes, develop slowly and may remain active until autumn frost.[378]

In the *Acer saccharum* Province three Sections may be distinguished. Across the cool northern border, climatic climaxes from northeastern Wisconsin[382] to the St. Lawrence Valley[94] are dominated by various combinations of *Acer saccharum, Betula alleghaniensis, Fagus grandifolia,* and the evergreen *Tsuga canadensis* (Fig. 51). *Oxalis montana, Taxus canadensis,* and *Viburnum alnifolium* are characteristic members of the undergrowth here. This Section also extends southward along the flanks of the Appalachian Mountains, especially on the western side, where taxonomic diversity increases markedly.[77,265,328] Deep, wide-bottomed valleys that extend well

Fig. 51.   Climatic climax forest of *Acer saccharum* with *Tsuga canadensis* and *Betula al-leghaniensis*, near Saranac Lake, New York.

back into the foothills of the southern Appalachian and Cumberland Mountains provide a combination of warmth and high rainfall from the approach effect. In such environments occur the most highly enriched variants of the *Acer–Fagus–Betula–Tsuga* group of forests, along with a considerable number of relics. In these southern montane extensions evergreen shrubs are often conspicuous in the undergrowth (Fig. 52).

Most of the coniferous elements of the *Acer saccharum* Province are confined to this Section. In the Lake States area *Thuja occidentalis* terminates

Fig. 52.   *Rhododendron maximum* and *Kalmia latifolia* forming a dense evergreen understory about 3 m tall beneath deciduous trees in winter condition. Foothills of the southern Appalachian Mountains in Tennessee.

succession on certain peatlands,[151] with a special ecotype seral in old pastures.[168] *Pinus strobus* and *P. resinosa* are seral trees following fire, with the former more restricted to zonal soils. Both occur in the Lake States, but only the former extends down the length of the Appalachians, as does *Tsuga canadensis*, where it joins *T. caroliniana*. *Taxus canadensis* is abundant only across the northern margin of the Section.

Zonal soils in this Section belong mainly to the Podzol and Gray–Brown Podzolic Groups, and it is the resinous foliage and litter associated with the conifers that make the forests of Podzol areas especially flammable. When Caucasians first started exploiting the area it had an abundance of seral pine forests, bespeaking the former frequence of holocausts. Among other seral plants of deforested land *Pteridium aquilinum* and trees such as *Alnus incana*, *Betula alleghaniensis*, *B. papyrifera*, *B. populifolia*, *Populus tremuloides*, *Prunus pennsylvanica*, and *P. virginiana* are conspicuous, especially in recent decades since the pine stands have been largely logged off and seed sources eliminated[329] (Fig. 53).

From Minnesota to New York large tracts of sandy glacial outwash support edaphic climaxes. Here *Pinus banksiana* is a major species and a glacial relic, but a few species of *Quercus* which tolerate the poor nutrition of these washed sands have been encorporated, or even assumed dominance in places. In eastern Massachusetts and New Jersey podsolized sands support *Pinus rigida* associated with *Quercus alba*, *Q. coccinea*, and *Q. velutina*.

Fig. 53. An abandoned field on stony glacial till in Massachusetts that was low in fertility even when first cleared of trees. In front of the stone fence the recently abandoned but still grazed field is dominated by *Rubus*. Beyond the fence an older field has grown up to a *Betula populifolia–Populus tremuloides–Pinus strobus* stand.

Immediately south of the western extremity of the above Section is another in which *Acer saccharum* and *Tilia americana* are the leading dominants of the climatic climaxes (Fig. 54). Forests of this type occur in a belt extending from Minnesota to southern Wisconsin and central Illinois. Southwest of this belt the type may be found sparingly to northern Missouri, but in this area it becomes limited to stream terraces and north-facing slopes which are relatively moist and so have been more protected from the occasional incursions of fire spreading from steppe on the west.[239] The seral trees of the *Acer–Fagus–Betula–Tsuga* Section do not occur here except as scattered glacial relics. Soils in this Section and the one to follow are mostly in the Brown Podzolic and Brown Forest Great Soil Groups.

In the third Section of the *Acer saccharum* Province, *Acer saccharum* and *Fagus grandifolia* are the characteristic climax dominants on zonal soils. Ohio, Indiana, and southern Michigan contain most of this Section. *Tilia americana* is uncommon in climax forests. *Betula alleghaniensis*, *Pinus strobus*, and *Tsuga canadensis* are present only as glacial relics restricted to very specialized habitats. Secondary succession here involves mainly abandoned fields, and some characteristic trees in this role are *Carya*, *Crataegus*, *Diospyros virginiana*, *Juniperus virginiana*, *Liquidambar styraciflua*, *Liriodendron tulipifera*, *Prunus virginiana*, *Quercus*, and *Sassafras albidum*.

Although the few trees mentioned in distinguishing the three Sections above are the species best represented in old, self-regenerating stands on zonal soils, there is also a rich assortment of species that maintain minor

Fig. 54. Interior of a virgin *Acer saccharum–Tilia americana* forest in winter condition, with a continuous carpet of leaves that were cast in the preceding autumn. Central Minnesota.

positions in the climaxes. Thus, all of these forests are distinctly more complex than those of the subarctic and subalpine forests of colder climates.

Special riparian forests in the *Acer saccharum* Province are dominated by *Acer saccharinum, Betula nigra, Celtis occidentalis, Fraxinus, Platanus occidentalis, Populus deltoides, Quercus palustris, Salix,* and *Ulmus. Acer rubrum* and *Thuja occidentalis* are especially common on the abundant areas of swampy ground, as are *Osmunda cinnamomum* and *Symplocarpus foetidus.*

Fen and bog communities closely similar to those indicated for the *Picea glauca* Province also occur in the glaciated portion of the *Acer* Province. These represent highly distinctive enclaves of boreal biota that became stranded early in post-Wisconsin time as the ecotone between the subarctic and temperate forest receded northward. Apparently the peat substrate loses its winter chill very slowly and so has not changed its character as much as has the soils of the surrounding mineral uplands where the flora responded so markedly to the climatic warming. However, *Picea mariana* stands often do not regenerate after deforestation, and a forest of *Acer rubrum, Fraxinus nigra,* and *Ulmus americana* then replaces them on the peat.

Along the Atlantic coast *Chamaecyparis thyoides* becomes a local bog dominant.

During the middle part of Holocene time, roughly between 8000 and 4000 BP, the effectivity of the Westerlies in dragging dry air from the western steppe eastward became much stronger than at present. This resulted in an eastward extension of grassland into what is now the southern part of the *Acer saccharum* Province, i.e., between the Ohio River and the Great Lakes (Fig. 55). At its zenith this "prairie peninsula" extended as far east as the foothills of the Appalachians.[393] Even today the climate along this track of the Westerlies is drier than areas immediately to the north and south, so not much of a reduction in the effectivity of summer precipitation would have been necessary to shift the competitive advantage from mesophytic trees to grasses.

As the Hypsithermal Epoch waned, forest started to reclaim the "prairie peninsula," but fires, undoubtedly of increasing frequency due to increasing aboriginal populations, were carried eastward by the Westerlies and periodically set back forest advances. When Caucasian man gained control of the landscape there were numerous fragments of forest isolated in the prairie peninsula where fire had offset westward advance, leaving forest relics to the lee of such fire barriers as marshes, ponds, and rivers. To the east of this fire-induced ecotone, which by then had shifted to approximately the western edge of Indiana (Fig. 55), there persisted many disjunct relics of grassland biota in the area previously overwhelmed by the climatically extended steppe. Relic colonies of *Bouteloua curtipendula* and the thirteen-lined ground squirrel persisting as far east as the Appalachians bear mute evidence of the approximate eastern terminus of the prairie peninsula (Fig. 56). Most of Illinois was still grassland two centuries ago, with the wedge

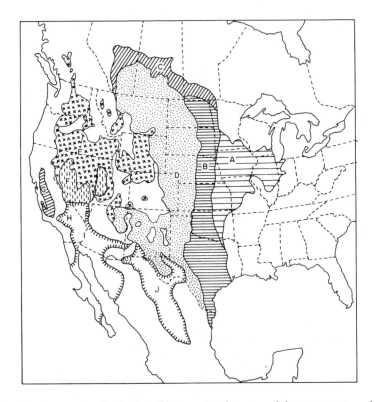

Fig. 55. Approximate distribution of major units of steppe and desert vegetation of North America, with the extension of steppe into Mexico mostly omitted owing to lack of consistency in existing vegetation maps. A, Fire-maintained grassland in forest climate; B, *Andropogon scoparius* Province; C, *Festuca scabrella* Province; D, *Bouteloua gracilis* Province; E, *Agropyron spicatum* Province; F, *Atriplex confertifolia* Province; G, *Stipa pulchra* Province; H, *Larrea divaricata* Province, Mojave Section; I, *Larrea divaricata* Province, Sonoran Section; J, *Larrea divaricata* Province, Chihuahuan Section.

broadening westward to include eastern Kansas and southern Minnesota (Fig. 57). Gallery forests in the prairie peninsula, especially on the leeward sides of rivers, included a rich assortment of species representing the forest to the east, including *Acer negundo*, *Fraxinus*, *Juglans nigra*, *Quercus rubra*, *Tilia americana*, *Carya* and *Ulmus* (Fig. 58). Wherever the early settlers protected the grassland from fire, forest spread rapidly from the riparian strips and other fire-protected refugia.

Since *Quercus macrocarpa* tolerates fire after the trunk develops a thick bark, and its seedlings are killed back only to the ground surface, this tree tended to border the fire-maintained grassland. It occurred as a savanna, with grasses typical of the steppe immediately west covering the ground quite completely (Fig. 59).

Fig. 56. *Quercus macrocarpa–Rosa acicularis* community, possibly a Hypsithermal relic, confined to a south-facing slope 29 km west of Ft. William, Ontario.

Fig. 57. Fire climax dominated by *Andropogon scoparius* and *A. gerardi* in southern Minnesota.

Fig. 58. Ravines filled with deciduous trees (*Celtis, Gleditsia, Quercus, Ulmus*) in the Flint Hills of eastern Kansas, where *Andropogon*-dominated grassland on the uplands has been maintained by frequent fires.

Andropogon gerardi was the major dominant of these fire-maintained grasslands, with *Panicum virgatum* and *Sorghastrum nutans* also well represented. *Andropogon scoparium* was limited to sandy or other dry habitats. Since there was sufficient precipitation to support forest, the soils became leached, and although blackened by humus they are classed as Prairie Podzolic.

Fig. 59. Savanna dominated by *Quercus macrocarpa* along the western edge of temperate forest in Minnesota.

Elsewhere in the *Acer saccharum* Province, on narrow ridgetops and south-facing slopes there are topographic climaxes with *Quercus* and *Carya* as major trees. These somewhat xerophytic topographic climaxes have close floristic and environmental affinities with the *Quercus falcata* Province to the south, and cover more and more of the topography southward across the *Acer saccharum* Province. Everywhere these species spread from their warm habitats onto cooler and wetter areas as seral dominants following deforestation.

Dunes along the Atlantic coast have *Ammophila breviligulata* as their most characteristic plant. Coastal salt marshes along this Province include *Distichlis spicata, Juncus gerardi, Spartina alternifolia,* and *S. patens.*

Along the northern edge of the *Acer saccharum* Province where it forms an ecotone with the *Picea glauca* Province, there are hybrid associations or alternations of essentially pure communities representative of the two Provinces, as described previously. In the Appalachians there is an ecotone with the *Picea rubens* Province, and along the southern border there is extensive alternation with communities typical of the *Quercus falcata* Province.

Some characteristic vertebrates of the primeval landscape included woodland bison, Virginia deer, mule deer, marmot, cottontail rabbit, muskrat, squirrel, chipmunk, black bear, raccoon, striped skunk, opossum, puma, bobcat, timber wolf, coyote, gray fox, red fox, weasel, mink, copperhead snake, garter snake, bluejay, brown thrasher, cardinal, grackle, turkey, and the extinct passenger pigeon.

A number of trees in this Province have valuable wood. *Juglans nigra, Liriodendron tulipifera, Pinus strobus, Prunus serotina,* and *Quercus alba* are particularly outstanding in this respect, although *Acer saccharum, Carya,* and *Fraxinus* are highly valued also. The *Acer* has long been a source of the highly prized maple syrup, and the fruits of *Carya, Fagus,* and *Quercus* provide mast for both domestic pigs and wildlife. *Castanea dentata* wood and nuts were both economically important before the species was essentially exterminated, and the bark was an important source of tannin.

Heavy grazing in this Province, mostly by milk cows, typically results in the development of a *Poa pratensis*-dominated zootic climax.

A rich diversity of crops are grown in the Province: tree fruits (apples, peaches, pears, cherries), berries (blackberries, gooseberries, grapes, raspberries, strawberries), vegetables (beans, cabbage, onions, peas, tomatoes, white potato), melons (muskmelon, watermelon), and field crops (hay, maize, oats, soybeans, tobacco, wheat).

The *Quercus falcata* Province

East of the Appalachians the *Quercus falcata* Province extends from New Jersey to far down peninsular Florida, and west of those mountains from Kentucky and Missouri to the Gulf of Mexico, including eastern Texas. Thus

it surrounds intrusions of the *Acer saccharum* and *Picea rubens* Provinces
that occur on the slopes of the Appalachians. The summers are longer and the
winters especially warmer than in the *Acer saccharum* Province (compare
Figs 11J,I), and the zonal soils are mainly Red–Yellow Podzolic.

The Province is complex, long subject to disturbance, and inadequately
studied. *Quercus falcata* extends essentially throughout the three Sections
recognized (Fig. 60).

Climatic climaxes in the northern Section, which includes all of the Province inland from the Atlantic and Gulf Coastal Plains, are dominated almost
exclusively by deciduous trees. *Quercus* and *Carya* are conspicuous among
these (Fig. 61), but a wide assortment of species in other genera are also well
represented.[327] *Castanea dentata* was especially an important component,
mainly on the foothills of the Appalachians. The highly discontinuous lower
tree stratum includes those listed for the *Acer saccharum* Province, plus
Acer barbatum, *Ilex opaca*, *Magnolia fraseri*, *M. tripetala*, *Oxydendrum
arboreum*, *Sassafras albidum*, and *Vaccinium arboreum*. The liana flora is
likewise enriched, notably by thorny species of *Smilax*. Among the rich
undergrowth flora of shrubs and herbs in this Province the small bamboo
Arundinaria gigantea (rarely to 10 m tall) determined the undergrowth
physiognomy rather extensively until it was virtually eliminated by cattle
grazing.

Aside from the prevalence of deciduous species among the climax trees,
another major feature distinguishing this Section is that in secondary succession *Pinus echinata*, *P. rigida*, *P. taeda*, and *P. virginiana* are conspicuous

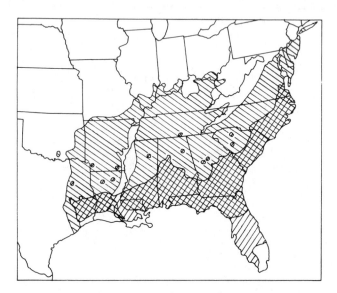

Fig. 60.   Ranges of the deciduous *Quercus falcata* (northerly) and *Q. laurifolia* (southerly).

Fig. 61.   *Quercus alba* in the winter aspect of a *Quercus–Carya* forest near Knoxville, Tennessee. The heavy leaf litter was cast during a brief period in autumn a few months previously.

seral dominants in various parts of the Section. Since these pines are not as sensitive to fire as are the dicot trees, frequent burning of the vegetation has maintained them as dominants in places.[253] *Andropogon virginicus* is a common perennial grass that precedes the pines in secondary succession and tends to persist in fire-maintained stands (Figs. 62 and 63).

In eastern Massachusetts and New Jersey podzolized sands support rather stunted forest of *Pinus rigida* associated with *Quercus alba*, *Q. coccinea*, and *Q. velutina*. Over a sizeable part of the area this vegetation exists as a frequently burned scrub less than 2 m tall (Fig. 64).[268] *Pinus rigida* also persists on dry stony ridges where it meets very little competition from other trees.

A vegetation catena of special interest is found on the sides or summits of mountain ridges and peaks south of the limits of the *Picea rubens* Province. Here there are islandlike areas of grassland dominated by *Danthonia compressa*, *Carex*, and *Potentilla canadensis*, which are surrounded, in turn, by belts of *Vaccinium* and other low deciduous shrubs, low trees, chief among which is *Crataegus*, then by an open and stunted belt of forest consisting of *Quercus rubra* with limited amounts of *Fagus grandifolia* and *Aesculus octandra* (Fig. 65). This life-form gradient, including the dwarfing of the forest margin, combined with meteorologic evidence of a decline in precipitation with increasing altitude,[112] suggest that these parks may be determined by dryness that may be aggrevated possibly by soil conditions. However, hypotheses to account for this vegetation catena are numerous.[275] Similar

Fig. 62.   Abandoned field with dominance passing from *Andropogon virginicus* to *Juniperus virginiana*, *Rhus glabra*, *Liquidambar*, and other broad-leaved trees. Arkansas.

Fig. 63.   In places the introduced subtropical *Pueraria thunbergiana*, a leguminous liana introduced to control erosion and provide forage on deforested land, spreads rampantly and overwhelms forests. Yazoo, Mississippi.

Fig. 64. A pyro-edaphic climax of *Pinus rigida*–*Quercus marilandica*–*Q. ilicifolia* that has been maintained on infertile sand by frequent burning.[268] All dominants sprout readily from the summit of their tap roots following fire. Southern New Jersey.

Fig. 65. *Danthonia compressa*–*Carex flexuosa* grassland in foreground, with a fringing scrub of *Vaccinium*, *Lyonia*, *Salix*, and *Crataegus* next the edge of dwarfed *Quercus rubra* forest. Gregory Bald. Great Smoky Mountain National Park, Tennessee.

parks occur as far north as Virginia as interruptions in the *Picea rubens* forests, but these are seral, at least locally,[52] suggesting their interpretation as Hypsithermal relics. The introduced European wild boar is especially damaging to the grassy vegetation of these parks.

Another special vegetation type in the Appalachian Mountains occurs on dry slopes and ridges where thin, noncalcareous soils support open xerophytic forests of mainly *Pinus pungens, P. rigida,* or *P. virginiana,* with a few broad-leaved trees and an ericaceous ground cover.

Sizeable areas on the basal plain west of the Appalachians have limestone or dolomite outcropping so near the soil surface that a highly distinctive vegetation is associated. Where the soil is very thin over these rocks, the original vegetation is suspected of having been an open forest or savanna with *Elymus* os possibly *Arundinaria gigantea* as a major component of the matrix. These areas, especially well represented in Kentucky, are now dominated by *Poa pratensis* and used mainly for pasture. Where the limestone is nearly bare *Juniperus virginiana* is the chief arborescent member of savannas which have such a sparse xerophytic vegetation that they have been essentially immune to alteration by either grazing or fire.[240,326]

In Alabama and Mississippi areas of heavy clay originally supported parks dominated by much the same grasses as occur in the eastern margin of steppe to the west of the Province, especially *Andropogon gerardi, Panicum virgatum,* and *Sorghastrum nutans,* together with an abundance of the forbs usually associated with them in the steppe.

North-facing slopes and other habitats with above-average moisture, yet with well-drained soils, support vegetation that represents outliers of the *Acer saccharum* series. These occur essentially throughout the Section under consideration.

On approaching the Mississippi River and other large streams, there are extensive lowlands with series of terraces, floodplains, and oxbow ponds, all of which support distinctive edaphic climaxes. Where the water table is very near the soil surface, or rises above the soil surface seasonally, variations in its height and duration are crucial factors differentiating associations composed variously of *Acer rubrum, A. saccharinum, Betula nigra, Carya aquatica, Celtis laevigata, Fraxinus pennsylvanica, Liquidambar styraciflua, Nyssa biflora, Platanus occidentalis, Quercus lyrata, Salix,* and *Ulmus americana* (Fig. 66). The lower tree layer here includes *Aralia spinosa* and *Asimina triloba* as common widespread members. Where water along river margins or circumneutral pond margins covers the surface usually for years at a time, there occur swamp forests composed almost entirely of *Taxodium distichum* with *Nyssa aquatica* or *N. sylvatica biflora* or *N. ogeche.* Here the bases of the trunks of both genera are conspicuously swollen, and curious upward growths ("knees") on the shallow lateral roots of *Taxodium* rise as high as the water surface at its highest stages (Fig. 67).

Fig. 66.   Luxuriant forest where the water table is not far below the surface. Barge Lake, Mississippi.

A central Section of the *Quercus falcata* Province embraces most of the Atlantic and Gulf Coastal Plains from North Carolina to eastern Texas. The aborigines here practiced shifting cultivation to take advantage of the fertility which natural vegetation accumulated over a number of years between successive cropping of the same land. Common field weeds throughout the *Quercus falcata* Province are perennial grasses which have a dry-season rest period, and the high heat level coupled with the sandy soils of this coastal plain favored fires that tended to perpetuate grass against the encroachment of dicot forest after land abandonment. Pines, especially *Pinus palustris*, prosper and reproduce where fires occurring at intervals of 3–10 years keep out superior competitors (Fig. 68). Thus, for centuries fire climaxes of pine with grass-dominated undergrowth (Fig. 69) have been the most prevalent types of vegetation in this Section. The grasses (*Andropogon, Aristida stricta, Panicum, Sporobolus curtisii*, etc.) provide cattle forage of only fair quality, but it is definitely superior to the dicots of climax or near-climax

Fig. 67.   *Taxodium distichum* swamp. Northern Florida.

Fig. 68.   *Pinus palustris*, with stump sprouts of *Quercus* spp., four years after the last surface fire. The pine seedling to the right of the meter stake is now tall enough to withstand another surface fire which would kill the root sprouts of the angiosperm trees back to the ground. Ocala, Florida.

Fig. 69. *Pinus palustris–Aristida stricta* forest, a product of frequent fire in a habitat that would otherwise be occupied by an angiosperm forest. Near Tallahassee, Florida.

forest. Since pine timber is more valuable than that from the dicot trees, this area lends itself well to management for combined grazing and lumbering. Other benefits from burning include improvement of habitat for wildlife (deer and quail) and a reduction of a disease which attacks pine seedlings. Much study has been devoted to make reasonably certain that no long-term deterioration of environment is necessarily associated with fire-based management here.

Although grass characterizes the undergrowth in large areas of periodically burned pine forest, other habitats, especially ground-water podzols, favor a shrubby fan palm (*Serenoa repens*) which forms a very dense but valueless undergrowth about 1.5 m tall (Fig. 70).

Owing to the long history of vegetation fires throughout this Section, very little is known about the climatic climax vegetation. However, small widely scattered places where marshes, swamps, or rivers have offered enough fire protection, islandlike stands of old dicot forests may be found on the uplands (Fig. 71). These stands, along with land survey records, provide some evidence of the the nature of the potential natural vegetation of unburned zonal soils.[39,113,178,244] The characteristic dominants of most such areas appear to have included a distinctive southern ecotype of *Fagus grandifolia* and the evergreen *Magnolia grandiflora*. Evergreen species of *Ilex* were usually major contributors to the lower tree stratum.

The southern Section of the Province extends from the base of the Florida peninsula to near its tip. In this latitudinal belt the climate begins to be clearly influenced by the Trade Winds. Air movement is predominantly from the east, and thunderstorms interrupt otherwise still air in summers. Win-

Fig. 70. *Pinus elliottii–Serenoa repens* forest, a fire climax in the Ocala National Forest, Florida.

Fig. 71. Interior of a *Fagus grandifolia–Magnolia grandiflora* forest near Tallahassee, Florida.

ters bring clear weather with breezes, but little rainfall. Owing to a reduction in the amount of freezing weather, the proportion of evergreen trees in the forest increases southward across the *Quercus falcata* Province. In this Section it may rise as high as 75%.

The oldest forests to be found on the sandy loams of uplands support *Quercus virginiana* and/or *Q. laurifolia* (Fig. 72), with the shade-tolerant alien *Cinnamomum camphora* now invading many of the stands. Of these three only *Q. laurifolia* is deciduous. Fire again favors *Pinus palustris*, with *Quercus laevis* also playing an important seral role. In this Section *Magnolia grandiflora* becomes restricted to floristically richer forests of moist lowlands, along with *Quercus virginiana*, *Sabal palmetto*, *Liquidambar styraciflua*, *Persea borbonia*, *Bumelia tenax*, *Morus rubra*, *Tilia floridana*, etc.

Where sandiness is more pronounced, infertile and droughty substrates of the uplands often support forests of *Quercus laevis* (deciduous) which, owing to the frequency of surface fires running through the deciduous leaf cast, remain open enough for the taller *Pinus palustris* to maintain a conspicuous presence. *Aristida stricta* is a common dominant of the ground cover in these distinctive open forests, which extend deeply into the central Section as well. Ancient dunes of sand so strongly sorted and leached as to be white bear a cover of tall scrub or woodland of *Quercus myrtifolia* and *Q. chapmani*, with *Pinus clausa* conspicuous as a seral tree.[246,247] This community is nonflammable to a remarkable degree, owing to its prevailing evergreen character, sparse litter, and near absence of herbs.

Fig. 72. *Quercus laurifolia–Q. virginiana* forest festooned with *Tillandsia usneoides*. Near Green Cove Springs, Florida.

Many ponds and depressions occur in this Section, as results of the caving-in of channels dissolved in limestone by subterranean streams. Seasonally wet soils support marshes or swamps consisting variously of *Acer rubrum, Carya aquatica, Cyrilla racemiflora, Gordonia lasianthus, Ilex, Magnolia virginiana, Nyssa sylvatica biflora, Myrica cerifera, Persea borbonia, Quercus nigra, Sabal palmetto.*[92,292] On gleyed soils *Pinus elliottii* occurred as an overstory above some of the foregoing woody plants. Where clay in the subsoil maintains ponds most of the time and the water is distinctly acid, *Taxodium ascendens* and/or *Nyssa sylvatica biflora* dominate swamps.[294] In other habitats where water stands on the surface only briefly, *Pinus serotina* is especially well represented, and here a dwarf bamboo, *Arundinaria tecta,* is a characteristic member of the undergrowth.

Coastal dunes around the *Quercus falcata* Province have *Uniola paniculata* as their most characteristic sand-binder. In the salt marshes behind these dunes the halophytes include *Distichlis spicata, Juncus roemerianus, Spartina alterniflora,* and *S. patens.*

In the southern part of this southern Section citrus fruits, sugar cane, vegetables, and some pineapples are grown on soils that are either sands or muck. To the north, and especially in the central Section where loams prevail, maize, cotton, rice, tobacco, sweet potato, peanuts, and crimson clover are major crops. Throughout fertilizer requirements are high.

Characteristic land vertebrates of the southern Section include white-tailed deer, gray squirrel, armadillo, black bear, bobcat, puma, and fox, and in the rivers alligator and manatee.

Southward across the *Quercus falcata* Province there is a progressive reduction in the amount of freezing weather, and on peninsular Florida summer rainfall begins to exceed that of winter. These trends culminate in tropical xerophytic forest climate at the southern tip of the peninsula as frost becomes a rare phenomenon. Not only does the percentage of evergreen species increase along this climatic gradient, but palms appear, first the arborescent *Sabal palmetto* then the shrubby *Serenoa repens,* then others. Vascular epiphytes also increase, starting with *Polypodium vulgare,* then *Tillandsia usneoides* and *T. recurvata* are added, with other bromeliads, ferns, and orchids enriching this synusia progressively southward.

At the western border of the Province where it meets the midcontinental steppe in Texas and Oklahoma, species of *Quercus,* especially *Q. marilandica* and *Q. stellata,* form savannas physiognomically identical with those formed by *Q. macrocarpa* along the steppe border to the north (Fig. 73). Also as in the north, *Symphoricarpos* and *Rhus glabra* are common shrubs in the forest–steppe ecotone. The grasses in these ecotonal savannas include *Andropogon, Panicum, Sorghastrum nutans, Tridens,* and *Uniola.* Broad tongues of such vegetation penetrate far out into the Steppe Region on sandy soils[122] with groves of shrub-size *Quercus harvardii* reaching as far as the eastern margin of New Mexico.

Fig. 73. *Quercus marilandica–Q. stellata* savanna along the western edge of temperate mesophytic forest near Henrietta, Texas.

The *Tsuga heterophylla* Province

As late as mid-Miocene time the western segment of the temperate mesophytic forest that had become separated from the eastern segment, was still floristically diversified over wide area. Along the Pacific coast it extended to Alaska and across Beringia, thus retaining continuity with the Asian segment.[437] But late in this Epoch summer temperatures appear to have dropped by about 7°C causing the northerly limits of the deciduous elements to retract to approximately central British Columbia, leaving the coniferous elements in control in Alaska.

In Pliocene time further restrictions came as a result of the accumulation of a great ridge of volcanic debris that formed the Cascade Mountains and the uplift that produced the Sierra Nevada Mountains to the south. These mountains formed a continuous chain which cut off so much precipitation to their leeward that a great arid trough developed between them and the Rockies. This nearly eliminated temperate mesophytic forest in the interior. Along the coast the climate was too dry for this forest to recede farther southward than approximately central California. In addition to this drastic reduction in area remaining suitably moist, qualitative changes in climate effected compositional alterations. Summers became cooler so that heat became inadequate for most deciduous trees to complete their normal summer activities before autumn frost, with a result that most of these trees died out leaving evergreen conifers in control. (This was a purge similar to that in eastern North America which allowed the evergreen conifer constituent to retain a place in only the higher altitudinal or latitudinal margin of temperate mesophytic forest, and similar to the segregation of evergreen species in the Alaskan highlands earlier.) The evergreens probably persisted because frost-free intervals in spring and autumn enabled them to photosynthesize during the many warm days when deciduous trees stood leafless, thus compensating for the coolness of summer. The few deciduous elements which were able to cope with this reduction of heat became subordinate to the conifers. Others (e.g., *Cercis, Dirca, Juglans, Ostrya, Platanus, Ptelea, Vitis*) took refuge far to the south in riparian habitats where their heat requirements could be met. The trees that became extinct in the west, but persisted in the eastern part of the continent, nearly all of them deciduous, include

*Acer(rubrum)**	*Ilex(opaca)*	*Quercus (falcata)*
A. (saccharinum)	*Liquidambar (styraciflua)*	*Sassafras (albidum)*
Betula(alleghaniensis)	*Nyssa (aquatica)*	*Taxodium (distichum)*
Diospyros (virginiana)	*Ostrya (virginiana)*	*Tilia*
Fagus (grandifolia)	*Pinus (echinata)*	*Ulmus (rubra)*
Gymnocladus (dioica)	*P. (taeda)*	
Hamamelis (virginiana)	*Ptelea (trifoliata)*	

Of the woody plants that persisted as vicariads in both segments, most are evergreen:

Acer macrophyllum—A. saccharum	*Quercus garryana—Q. alba*
†*Chamaecyparis lawsoniana—C. thyoides*	†*Rhododendron macrophylla—R.*
Cornus nuttallii—C. florida	*catawbiense*
Fraxinus latifolia—F. americana	†*Taxus brevifolia—T. floridana*
†*Pachistima myrsinites—P. canbyi*	†*Thuja plicata—T. occidentalis*
†*Pinus monticola—P. strobus*	†*Tsuga heterophylla—T. canadensis*
†*Polystichum munitum—P. acrostichoides*	†*Xerophyllum tenax—X. asphodeloides*

Relatively few, all evergreen, persisted only in the western segment:

Abies (grandis or *concolor)*	*Pseudotsuga*
Libocedrus	*Sequoia*
Berberis (Mahonia section)	*Sequoiadendron*

All these restrictions in area and reductions in floristic diversity in the west were essentially completed before the start of Pleistocene glaciations.

As the western forest evolved, it differentiated into what may be considered two Provinces. One of these, the *Tsuga heterophylla* Province, extends from Cook Inlet, Alaska, to central California. It is largely confined to a narrow strip between the Pacific Ocean and moderate elevations in the first major mountain ranges to the east, plus an enclave that persisted inland where the storm track of the Westerlies carries oceanic influences as far as the divide of the Rockies. The other Province, the *Pseudotsuga menziesii* Province, is more widely distributed but in drier climates, occurring mainly

---

*Names in parentheses indicate nearest living relatives as judged from macrofossils.[67]
†Evergreen.

at midaltitudes in the mountains where the heat supply is better than in subalpine forest, yet moisture is not as deficient as on the arid lowlands.

Although temperate in their broader aspects, both of these Provinces lack the warm, hot summers of the eastern Provinces. The summers in the west are either cool, as along the coast, or subject to occasional frost and rather droughty, as on the slopes of the interior mountains. It is convenient to discuss separately three Sections of the *Tsuga heterophylla* Province.

The Province as a whole is the most luxuriant conifer forest in the world, containing the largest and most long-lived species in several genera: *Abies, Chamaecyparis, Larix, Libocedrus, Picea, Pseudotsuga, Sequoia, Thuja,* and *Tsuga.* These forests also have the greatest phytomass of any, probably well exceeding 2000 mT/hectare. Obviously the gene pools of these plants were not significantly depleted during the stresses of Pleistocene glaciations, and the survivors have been able to take full advantage of the minimal temperature and moisture stresses of the oceanic climates. None of these superlatives can be applied to the homologous vegetation of coastal Chile, which is dominated by *Nothofagus (Fagaceae)* and *Araucaria (Coniferae).*

The northern Section of the *Tsuga heterophylla* Province, extending from Cook Inlet to approximately the southern tip of Alaska, is floristically quite simple.[421] *Tsuga heterophylla* is the outstanding climax dominant of zonal soils from sea level to the subalpine *Tsuga mertensiana* forests, with *Picea sitchensis* the only major seral tree.[99] Minor trees include *Chamaecyparis nootkatensis* and *Tsuga mertensiana* which, though dominant in the subalpine forests above, extend to low altitude in the tracts of cold air continually drifting down valleys from the nearby glacier-clad mountains (Fig. 11K). Undergrowth plants include *Cornus canadensis, Menziesia ferruginea, Rubus pedatus, Vaccinium membranaceum,* and *V. ovalifolium,* these scattered over a continuous carpet of mosses such as *Calliergonella schreberi, Hylocomium splendens, Rhytidiadelphus loreus,* and *Rhytidiopsis robusta.*

Angiosperm trees are represented by *Alnus rubra* and *Populus trichocarpa,* which are seral on floodplains and young glacial moraines. The shrub *Alnus sinuata* is especially important in the latter role, and in addition persists in pure stands on snow-avalanche tracks.

Where the conifers grow on wet ground the thorny shrub *Oplopanax horridum* forms a conspicuous stratum, with *Lysichitum americanum* characteristic of swamps. Raised bogs are abundant, these supporting *Empetrum nigrum, Kalmia latifolia* and *Vaccinium oxycoccus* on the *Sphagnum* mat.[343] A scattering of dwarfed *Pinus contorta* (or *Chamaecyparis* or *Tsuga*) is found on some of these bogs. The *Pinus* also occurs to a limited extent on rock outcrops.

Sphagnum has been reported as spreading locally to form blanket bogs on uplands, replacing forest by preventing seedling establishment.[383,442] If continued this would end in treeless blanket bog homologous with that of the west coast of Ireland at similar latitudes.

The extensive strip of *Calluna vulgaris* scrub along the west coast of Europe from France to Norway, is generally attributed to deforestation followed by repeated burning of the vegetation, which has favored ericaceous shrubs and promoted the accumulation of a thick duff as well as ortstein in the soil. This tendency has not yet been recognized in the corresponding strip of oceanic climates along the west coast of North America.

At the northern extremity of this Section *Picea sitchensis* extends beyond the limits of all other conifers and so achieves climax status. The tree has reached this position only in recent decades, and is advancing rapidly onto the margin of adjacent tundra.[162]

The central Section of the *Tsuga heterophylla* Province extends from the southern tip of Alaska to approximately the southern edge of Oregon.[139,226,252] It occurs from sea level to midslopes of the Coast Ranges and Cascade Mountains, and a relictual detached portion, centered on the storm track of the Westerlies, extends inland as far as the continental divide in Montana and British Columbia.

On the whole, this warmer Section (Fig. 11L) is unique for its floristic richness in the tree layer. The few deciduous angiosperm trees that persisted along the Pacific Coast are best represented here.

On zonal soils west of the Cascades the climatic climaxes are dominated by *Tsuga heterophylla* over most of the area, with *Abies grandis* replacing it as climates become too dry in some of the valleys. Seral trees include *Pseudotsuga menziesii*, the most ubiquitous invader of burned areas, *Alnus rubra*, second only to *Pseudotsuga*, *Thuja plicata*, widespread as a late-seral tree, *Picea sitchensis*, chiefly in a narrow belt near the ocean, *Acer macrophyllum*, in wetter sites, *Pinus contorta*, on sandy soils next the ocean, *Abies grandis*, along the warm-dry limits, *Populus trichocarpa* and *Fraxinus latifolia* on floodplains and low stream terraces, *Larix occidentalis*, and *Pinus monticola*. Major undergrowth species include *Polystichum munitum*, *Gaultheria shallon*, *Berberis nervosa*, *Oxalis oregana*, *Vaccinium*, and *Rhododendron macrophyllum*. The weakly defined lower tree stratum consists of *Acer circinatum*, *Cornus nuttallii*, and *Taxus brevifolia*. Bryophytes and lichens are abundant everywhere owing to the high rainfall and mild winters which allow the numerous evergreen species of the undergrowth, as well as the trees, to be active at least sporadically in winter. Zonal soils in these forests embrace Reddish–Brown Laterite, Western Brown, Gray–Brown Podzolic, Sols Bruns Acides, Gray Wooded, and Podzols.

Following fire *Pteridium aquilinum*, *Rubus*, and *Epilobium angustifolium* are usually conspicuous for a short time before *Pseudotsuga*, *Alnus rubra*, or other less abundant seral trees overtop them. In this Section *Pseudotsuga* forms the base of an important lumber industry. The first settlers found magnificent stands of this tree growing up to 117 m tall, with a diameter at breast height reaching 4.7 m, as a consequence of prehistoric holocausts (Fig. 74).

Fig. 74.   Seral *Pseudotsuga menziesii* in a *Tsuga heterophylla–Polystichum munitum* habitat type. The fern, *Polystichum*, dominates a type of undergrowth that indicates the highest site productivity for *Pseudotsuga* in this area. Olympic Peninsula, Washington.

Swamps are usually dominated by *Thuja* or *Alnus rubra*, and have *Oplopanax horridum*, *Athyrium filix-foemina*, *Lysichitum americanum*, or *Blechnum spicant* as major undergrowth plants in the different associations (Fig. 75). Bogs are well represented in this Section, and the most common tree to make a limited invasion of their *Sphagnum* covering is *Pinus contorta*. In tidal marshes *Salicornia pacifica*, *Carex lyngbyei*, and *Distichlis spicata* provide most of the phytomass (Fig. 76). In the salt spray zone just above tide levels, salt-tolerant ecotypes of *Pinus contorta*, *P. sitchensis*, or *Alnus rubra* form pure stands, with their form often modified strongly by the salt spray.

Valley bottoms in the western foothills of the Olympic Mountains have high rainfall (to 3370 mm) and much fog owing to the approach effect on the Westerlies, and here epiphytic cryptogams, especially *Selaginella oregana*, are draped in profusion over the trunks and branches of *Acer macrophyllum* that forms a second tree layer beneath the taller *Tsuga* or *Picea sitchensis* (Fig. 77).[194]

Fig. 75. Swamp forests dominated by *Thuja plicata*, with a layer of *Oplopanax horridum* beneath, occur over most of the *Tsuga heterophylla* Province. The black tip of a meter stake shows at the lower right. Pend Oreille County, Washington.

On the other side of the Olympic Mountains, in the rain shadow, gravelly soils bore *Quercus garryana* savanna, with a ground cover of *Pteridium aquilinum* or grasses a meter tall, and here there were also grass-covered parks ringed with *Pinus ponderosa* when Caucasian man came upon the scene.

The vegetation mosaic characterized above extends from the coast up to the subalpine forests of the Cascades and Coast Ranges, except where inter-

Fig. 76. *Salicornia pacifica–Distichlis spicata–Atriplex–Cuscuta* marsh on a flat inundated by high tides. Seattle, Washington.

Fig. 77. *Acer macrophyllum* festooned with *Selaginella oregana* in the Quinault Valley on the windward side of the Olympic Mountains in Washington.

rupted by dry valleys west of the Cascades in Oregon. In this Section *Tsuga heterophylla* extends well up into the subalpine *Tsuga mertensiana* belt, but a reasonably good floristic break in associated species serves to distinguish temperate from subalpine vegetation. This ecotone is marked by the dropping out of *Acer macrophyllum* and *Alnus rubra*, and a reduction in the abundance of *Thuja*, that coincides fairly well with the appearance of *Abies procera, A. lasiocarpa*, and *Chamaecyparis nootkatensis* and a strong increase in the abundance of *Abies amabilis*. Then in the undergrowth *Gaultheria shallon* and *Polystichum munitum* give way to *Menziesia ferruginea, Rhododendron albiflorum*, and *Xerophyllum tenax* as common dominants.

The isolated segment of this Section that has persisted inland on the western slopes of the major mountain ranges astride the international border, differs in several ways from the segment along the coast.

Tsuga heterophylla, Thuja plicata, and *Abies grandis* each dominate climatic climaxes (Figs. 78 and 79) in successively drier environments.[109]

Fig. 78. Virgin forest with *Tsuga heterophylla* (right) and *Thuja plicata* (left) in northern Idaho.

Fig. 79. Climax forest dominated by *Abies grandis*. Harpster, Idaho.

The undergrowth here lacks *Gaultheria shallon* and *Oxalis oregana*, with *Polystichum munitum* and *Berberis nervosa* very poorly represented. Instead, a rich assortment of herbs and small shrubs, including *Clintonia uniflora*, *Pachistima myrsinites*, and *Tiarella unifoliata* form the undergrowth without any one species dominating. *Taxus brevifolia* is again common as a low tree (or shrub), but *Cornus nuttallii* is restricted to one small locality, and *Acer circinata* is lacking. *Pinus monticola, Larix occidentalis, Pseudotsuga menziesii*, and *Pinus contorta* are the major trees to invade burned land (Fig. 80) temporarily, but the inland ecotypes of *Pseudotsuga* are relatively small trees, and *Pinus monticola* was the leading timber species until introduced diseases reduced its economic status. The soils here range from Brown Podzolic to Gray Wooded to Gray–Brown Podzolic.

Inland swamp forests of *Thuja plicata* with *Oplopanax horridum* and *Athyrium filix-foemina* and riparian forests of *Populus trichocarpa* are much like those on the coast. Frost pockets in these mountains are marked by subalpine trees descending into these low montane forests (Fig. 81).

These inland representatives of the *Tsuga heterophylla* Province occur on intermediate mountain slopes, with the *Picea englemannii* Province above and the *Pseudotsuga menziesii* Province below.

The southernmost and warmest (see Fig. 12A) Section of the *Tsuga heterophylla* Province, extending along the seaward slope of the Coast

Fig. 80. Burned-over land where *Thuja plicata* is the potential climax dominant, now supporting a seral community of many species of shrubs which provide excellent winter browse for game herds for several decades after a fire. Many other forest types in the Rocky Mountains regenerate themselves directly after a fire without an intervening stage of deciduous shrubs.

Fig. 81. Fen dominated by *Carex* and dominated by a ring of *Abies lasiocarpa* which reflects frost-pocket influence, with *Thuja plicata* the climax dominant on higher ground. Clearwater Valley, Idaho.

Ranges from the southwest corner of Oregon to approximately central California, is distinguished by the dominance of *Sequoia sempervirens* (Fig. 82), an endemic that is restricted to this Section.[72] This tree attains a maximum height of 117 m, a diamter at breast height of 7 m, and an extreme age of about 2200 years. It is very valuable for its fine-grained, resin-free, rot-resistant, red-brown wood. On zonal soils of the slopes (Red–Yellow Podzolic, Brown Podzolic) this tree has an understory of scattered low trees, mostly *Lithocarpus densiflorus*, shrubs, especially *Vaccinium ovalifolium* and *V. parviflorum*, and forbs that include a sprinkling of *Oxalis oregana* and *Polystichum munitum*.

Although *Sequoia* is quite shade tolerant, the seedlings regularly succumb to fungi when they germinate on litter, duff, or even where there is much humus in the A horizon. However, owing to the longevity of the species, seedlings get established often enough on moss mats, rotting logs, and mineral material exposed by a fallen tree for the species to retain its dominance in the tallest stratum, with all other trees remaining much shorter. *Sequoia* is also remarkably tolerant of surface fires. The bark of old trees is quite thick and not very flammable, and during their first year seedlings develop lignotubers which allow replacement of damaged shoots. Even after successive enlargements of a basal fire scar causes an old tree to snap off, a circle of sprouts arises from dormant buds persisting just under the soil surface. The stump of a cut tree produces similar sprouts.

Tsuga heterophylla is the most aggressive seral tree in this Section, but requires disturbance, as from a surface fire, to stimulate a period of reproducing beneath the *Sequoia*. Locally, there are also individuals or small populations of *Thuja plicata*, *Abies grandis*, or *Chamaecyparis lawsoniana*,

Fig. 82. *Sequoia sempervirens* stand on a stream terrace. California.

most of which are restricted to low elevations on the slopes, and like *Tsuga*, appear to require minor disturbances to maintain their limited representation. *Pseudotsuga menziesii* is the principal seral tree following complete deforestation.

Sequoia forests of distinctive character occur as edaphic climax on floodplains and terraces where silt deposits exclude practically all other trees by root smothering. As its lower trunk becomes buried, *Sequoia* puts out adventitious roots into the alluvial blanket, with old roots turning their tips upward as they continue growth. The bases of living trees have been buried up to 9 m, having endured increments of silt almost a meter thick. Fresh silt is abundantly available here for new seedling establishment. Both *Picea sitchensis* and *Alnus rubra* are important seral trees on the floodplain and terrace sites along the lower stretches of the rivers.

Picea sitchensis and *Alnus rubra* also fringe the Province next the ocean in the northern part of the Section, for *Sequoia* is intolerant of salt spray. Southward a fringe of *Pinus muricata* forms open stands in this position, with a mostly ericaceous shrubby undergrowth.

Fifty to 100 meters above sea level there are ancient sandy beach deposits that have become strongly podzolized and support a scattering of very dwarfed trees—*Cupressus pygmaea, Pinus contorta bolanderi,* and *P. muricata*—with a thin lower stratum of ericaceous shrubs such as *Arctostaphylos, Gaultheria shallon, Ledum glandulosum,* and *Rhodendron macrophyllum.*[216]

High ridges exposed to strong winds support extensive parks dominated by *Calamagrostis nootkatensis, Danthonia californica, Deschampsia caespitosa,* and *Festuca idahoensis.* On windswept headlands along the coast *Baccharis pilularis* is a representative dominant of strongly contoured scrub.

As the Cold Polar air mass shrinks in summer, the Westerlies follow it northward, continuing to provide the northernmost Section of the *Tsuga heterophylla* Province with rainfall, but leaving the central and southern Sections rainless at this season (compare Fig. 11K with Figs. 11L and 12A). In the central Section the rainless season is too short to exhaust soil moisture seriously. In the southern Section the rainless season is longer and more absolute, but an abundance of cloud at this season, drifting off the ocean and extending up to 500 km inland before evaporating, compensates the lack of

Fig. 83. Near its southerly limits along the seaward face of the Coast Ranges, *Sequoia* becomes restricted to valleys that frequently, as here, fill with fog off the adjacent ocean. The rainfall is low, and strongly insolated ridges support only grassland.

rain by providing cool and humid air. At the northern end of the cloudy belt the *Sequoia* Section gives way to the central Section with *Tsuga heterophylla* the leading climax dominant. Inland, where the overcast ends, the ecotone is with the *Pseudotsuga menziesii* Province. The southerly limit of the *Sequoia* forests is related to the lengthening of the rainless summer, coupled with a reduction in the amount and frequency of fog or cloud. There forest is replaced spatially by steppe or chaparral, with the last groves of *Sequoia* becoming restricted to deep canyons (Fig. 83).

The *Pseudotsuga menziesii* Province

The *Tsuga heterophylla* Province represents a rather coherent and largely coastal remnant of the early Cenozoic temperate mesophytic forest which retained many floristic components of the ancestral assemblage despite the assumption of dominance by evergreen conifers. In less oceanic climates (compare Fig. 12B with Figs. 12A, 11K, L) and over a much greater area of the Cordillera, a floristically more impoverished remnant of that forest has

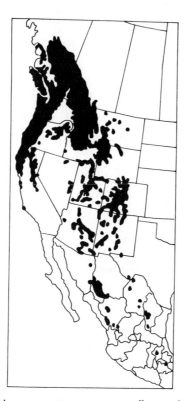

Fig. 84. Range of *Pseudotsuga menziesii*, an ecotypically very diversified species that plays seral or climax roles in a wide variety of habitat types.

persisted on intermediate mountain slopes, hemmed in between the xerophytic vegetation of the foothills (with Madro–Tertiary affinities) and forests of either the *Tsuga heterophylla* Province or subalpine forests above. This remnant is distinguished by usually having *Pseudotsuga menziesii* as an important climax dominant. Since it is spread over wide area (Fig. 84), there is Sectional differentiation.

A western Section occupies intermediate mountain slopes from south-western Oregon to northern Baja California, occurring extensively on the west slope of the Cascade and Sierra Nevada Mountains, and locally on the landward side of the Coast Ranges in northern California and southern Oregon (Fig. 85). Climax forests are composed variously of *Pseudotsuga menziesii* (mainly in the north), *Abies concolor*, *Libocedrus decurrens*, and *Pinus lambertiana*, with the latter two becoming more regular in occurrence southward (Fig. 86). Often a lower tree stratum is present, including such species as *Arbutus menziesii*, *Castanopsis chrysophylla*, *Cornus nuttallii*, *Lithocarpus densiflorus*, *Quercus chrysolepis*, *Q. kelloggii*, or *Taxus brevifolia*. The soils of these forests appear to belong mainly to the Reddish-Brown Lateritic or Red–Yellow Podozolic Groups.

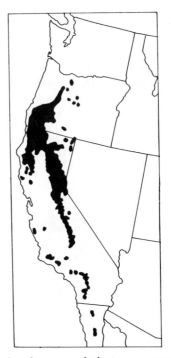

Fig. 85.   Range of *Libocedrus decurrens*, which gives an approximation of the western Section of the *Pseudotsuga menziesii* Province.

Fig. 86. *Pseudotsuga menziesii–Libocedrus decurrens–Pinus lambertiana–P. jeffreyi* forest with evergreen shrubs in the undergrowth. Rogue River National Forest, Oregon.

Two paleoendemics of special interest occur in this Section—*Picea breweriana* and *Sequoiadendron gigantea,* the former limited to an area centered on the northwest corner of California, the latter restricted to valleys on the west slope of the Sierra Nevada. *Sequoiadendron,* easily the more famous, is the most massive tree in North America, growing to 100 m tall and over 10 m in diameter at breast height (Fig. 87). It is among the

Fig. 87. *Sequoiadendron gigantea* dwarfs the associated *Abies concolor* and *Libocedrus decurrens* which would eventually eliminate the *Sequoiadendron* (after its 3000-year life span!) in the absence of fire.

oldest living organisms, with individuals exhibiting up to 3214 xylem layers[354]—an age exceeded only by *Pinus longaeva*. This shade-intolerant tree owes its persistence on zonal soils to its extreme longevity, to fire-resistant bark that may be up to 75 cm thick, and to frequent surface fires that have thinned the climax dominants sufficiently to allow seedlings to get established from time to time. The 32 disjunct populations of *Sequoiadendron* occur over an area 400 km long, but cover only about 130 ha in the aggregate, so the tree is more notable for its remarkable size and age than as a widespread member of the *Pseudotsuga* Province.

The undergrowth in this Section of the Province includes *Amelanchier, Chamaebatiaria millefolium, Fremontia californica, Goodyera, Prunus, Pyrola picta, Rhus diversiloba, Ribes,* and *Symphoricarpos,* along with some fire-favored evergreen shrubs ecologically related to if not identical with the chaparral at lower altitudes: *Arctostaphylos, Castanopsis sempervirens, Ceanothus,* and *Quercus.*

Pinus jeffreyi is a relatively fire-resistant seral tree which, like *Sequoiadendron,* persists indefinitely wherever surface fires are frequent enough to keep the forest open. *Quercus kelloggii,* a small deciduous tree, is also greatly favored by fire in the lower part of the Section. At high altitudes *Pinus contorta* invades burns temporarily.

Where serpentine soils occur in the Section the normal climax trees often become scattered and stunted, and *Pinus jeffreyi* can maintain a place in such an open edaphic climax (Fig. 88).

Owing to the summer-dry climate, surface fires have been so frequent in this Section that most of the existing old forests are best considered as a

Fig. 88. *Pinus jeffreyi* and *Arctostaphylos* on a serpentine outcrop near Selma, Oregon.

weakly differentiated fire-climax complex. In places, fires have been more devastating or more frequent and trees have been completely eliminated leaving evergreen shrubs in control (Fig. 89).[428]

Under cooler environment this Section gives way to *Abies magnifica* stands of the subalpine belt, and under drier environment to *Pinus ponderosa* stands of the temperate xerophytic group.

The eastern Section of the *Pseudotsuga menziesii* Province extends from southern British Columbia to northern Mexico, and from the rain shadow of the Coast Ranges in Oregon to central Montana. Typically the Section lies between temperate xerophytic forests of the foothills and subalpine forests of the high ridges, but in the latitudes where the Westerlies drag oceanic influences far inland it may be bounded above by the *Tsuga heterophylla* Province.

Pseudotsuga menziesii is the most common climax dominant of zonal soils throughout (Fig. 90). In the southern Rockies *Abies concolor* is a second common climax dominant, which is found sparingly as far north as southern Oregon and Idaho. In the same latitudes *Picea pungens* occurs as a mainly riparian tree. Undergrowth in these forests varies from essentially pure grass cover (e.g., *Calamagrostis rubescens*) to low scrub (e.g., *Symphoricarpos*) to tall scrub (e.g., *Arctostaphylos nevadensis* or *Physocarpus malvaceus*). *Arnica cordifolia* is an especially characteristic forb. Soils range from Chernozem to Prairie, Gray–Brown Podzolic, Gray Wooded, and Podzol.

All three of the trees mentioned above are easily killed by fire, and over wide area their places are temporarily usurped by *Pinus contorta*, *P. ponderosa*, or *Populus tremuloides* (Fig. 91). To these widespread fire-followers are added *Pinus strobiformis* in the southern Rockies, *Larix occidentalis*, in

Fig. 89. Evergreen shrubs have proliferated following deforestation. Shasta, California.

Fig. 90. Virgin climax stand of the *Pseudotsuga menziesii–Calamagrostis rubescens* association. East slope of the Cascade Mountains, Washington.

Fig. 91. *Abies concolor* dominating the understory in a seral stand of *Pinus ponderosa*, *Populus tremuloides*, and *Picea pungens*. Coyote, New Mexico.

Fig. 92.   Large *Larix occidentalis* that invaded after an earlier forest fire, has become surrounded by a dense stand of young *Pinus contorta* that became established after a second fire. *Pseudotsuga menziesii* the climax dominant of the habitat type, appears beneath the *Pinus.* North Fork of Flathead River, Montana.

the northern Rockies (Fig. 92), and *Quercus garryana* in the Willamette Valley of Oregon.

Large areas in the central Rockies, at elevations where *Pseudotsuga* might be expected to dominate, have nonzonal soils with pure stands of either *Pinus contorta*[290] or *Populus tremuloides*[193,334] which show every evidence of maintaining themselves (Figs. 93 and 94). Elsewhere the continuity of forest is interrupted by edaphically determined parks, which if well drained are usually grass covered east of the continental divide and usually *Artemisia* dominated west of the divide (Figs. 95 and 96). In the Willamette Valley some knobs with southwest exposure support *Quercus garryana* forest as a topographic climax, with the most extremely exposed spots treeless.

There is much interfingering with contiguous Provinces, with *Pseudotsuga* forests sending peninsular extensions down cool ravines into the relatively xerophytic belt below, and penetrating the subalpine belt above by way of southerly slopes and dry ridges (Fig. 97).

Fig. 93. *Pinus contorta–Purshia tridentata* association as an edaphic climax on coarse sand. Yellowstone National Park, Wyoming.

Animal life of these midmontane forests includes wapiti, mule deer, white-tailed deer, porcupine, beaver, red squirrel, black bear, puma, coyote, bobcat, otter, jay, grouse, and magpie.

Justification for considering the *Pseudotsuga* Province as part of the Temperate Mesophytic Forest Region is as follows: (1) *Pseudotsuga* is a genus that was a member of the Arcto–Tertiary Geoflora, and has never been a

Fig. 94. Climax stand of *Populus tremuloides* on the west slope of the Rocky Mountains. Yellowstone National Park, Wyoming.

Fig. 95. *Pinus contorta* surrounding a park dominated by *Festuca ovina* and *Poa* on the east slope of the Rocky Moutains. Medicine Bow National Forest, Wyoming.

Fig. 96. Park dominated by *Artemisia cana* and *Symphoricarpos*, surrounded by climax *Populus tremuloides* forest on the west slope of the Rocky Mountains, Wyoming.

Fig. 97. *Pseudotsuga menziesii–Physocarpus malvaceus* as an edaphic climax on the thin soil of an exposed ridge at relatively high altitude in northern Idaho. More mesophytic forests of *Thuja plicata* and *Tsuga heterophylla* occur just below the summit on either side.

component of the subarctic–subalpine series. (2) Its modern associates, *Abies concolor*, *Libocedrus*, *Picea pungens*, and most undergrowth plants are likewise derivatives of that Geoflora, or at least have similar climatic requirements. (3) *Pseudotsuga* forests fit into the general scheme of climate-vegetation relationships by occupying a geographic position between cold Subarctic–Subalpine and dry Xerophytic Forest Regions.

The major derivatives of the temperate mesophytic forest in North America have been considered above, and some generalizations can now be summarized. Just as there was a strong segregation of tropical and temperate elements that coexisted early in the Cenozoic Epoch, there has been almost as much segregation of coniferous and angiosperm elements within the temperate forest itself, with the coniferous element becoming restricted to cool–summer phases of temperate climates. Another striking aspect of Cenozoic history has been the cleavage of temperate forest into eastern and western areas separated by midcontinental aridity. As vicariads, some genera retained a measure of importance either in climax or seral communities in both eastern and western segments, e.g., *Chamaecyparis*, *Thuja*, *Tsuga*, *Pinus* (*strobus* and *monticola*). Others like *Oplopanax* and *Euonymus* have retained good representation in only one of the segments. Many genera, including *Pseudotsuga*, *Sequoia*, *Fagus*, and *Carya*, survived in only one of the segments. Still another pattern is evident in chains of vicariads anchored in the *Quercus falcata* Province on the east and usually extending to California on the west:

Cercis canadensis, reniformis, occidentalis
Fraxinus americana, texensis, velutina
Juglans nigra, rupestris, californica
Platanus occidentalis, wrightii, racemosa
Prunus virginiana virginiana, virginiana melanocarpa, virginiana demissa
Vitis aestivalis, arizonica, girdiana

The western members of these series are invariably minor components of the vegetation, and for the most part have become restricted to habitats of above-average moisture supplies in the warm-dry climates of low elevations. The southerly distribution of most of them but emphasizes the requirement of temperate zone angiosperms for summer warmth.

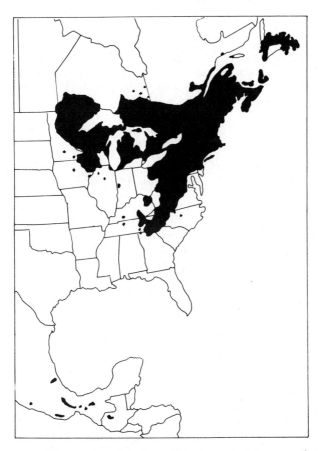

Fig. 98. Range of *Pinus strobus*, including variety *chiapensis* in Mexico and Guatemala.

FORESTS OF TEMPERATE AFFINITY IN MIDDLE
AMERICA[120,288]

On midmontane slopes well within the tropics, from the Mexican State of San Luis Potosi to Nicaragua, there occur mesophytic forests in which genera if not species primarily characteristic of the Temperate Mesophytic Forest Region are well represented (Fig. 98). At one station in this vegetation in Puebla the mean monthly temperatures are reported to range between 10–19°C, with an absolute minimum of −4°C.

As Cenozoic cooling forced plants to migrate southward, these temperate species were able to forge ahead of the main transcontinental belt, and ascended cool mountain slopes as they moved into tropical latitudes. According to fossil records, at least 10 genera of trees and shrubs with such affinities reached Veracruz in mid-Miocene time, shortly after a mountain axis up-

TABLE 4

Some Disjunct Species and Vicariads Derived from the Early Cenozoic Temperate Forest Flora that Extended from Coast to Coast Across the Northern Half of North America[a]

Atlantic North America	Pacific North America	Middle America
Acer saccharum	A. grandidentatum	A. skutchii
Adiantum pedatum	A. pedatum	A. pedatum
Botrychium virginianum	B. virginianum	B. virginianum
Ceanothus spp.	Ceanothus spp.	C. coeruleus
Chimaphila umbellata	C. umbellata	C. umbellata
Cornus florida	C. nuttallii	C. florida
Crataegus spp.	Crataegus spp.	C. pubescens
Fraxinus americana	F. latifolia	F. uhdei
Galium triflorum	G. triflorum	G. triflorum
Hypopitys monotropa	H. monotropa	H. monotropa
Juglans nigra	J. californica	J. pyriformis
Leucothoe catesbaei	L. davisiae	L. mexicana
Monotropa uniflora	M. uniflora	M. uniflora
Pentstemon spp.	Pentstemon spp.	P. campanulatus
Philadelphus spp.	P. lewisii	P. mexicana
Pinus strobus	P. monticola	P. strobus chiapensis
Platanus occidentalis	P. racemosa	P. chiapensis
Pterospora andromedea	P. andromedea	P. andromedea
Rhamnus alnifolia	R. alnifolia	R. caroliniana
Sambucus racemosa	S. racemosa	S. mexicana
Satureja spp.	S. douglasii	S. laevigata
Smilacina racemosa	S. racemosa	S. racemosa
Taxus floridana	T. brevifolia	T. globosa
Tiarella cordifolia	T. unifoliata	T. sp.
Triodanus perfoliata	T. perfoliata	T. perfoliata

[a] Similar patterns occur in bryophytes and in animals.

lifted to make the migration possible, with three of them (*Alnus, Juglans, Myrica*) having reached Panama by that time.[160] It is not surprising that this floristic element occurs on the older mountains in Guatemala, but is lacking on mountains of Pleistocene age.[386]

In the Pliocene Epoch, as aridity intensified in the subtropical high pressure belt behind them, these precocious taxa were cut off from the main body of the temperate flora which remained in temperate latitudes. Vertebrate geography also reflects this isolation of temperate zone organisms in the mountains of Mexico prior to the expansion of aridity in Pliocene time.[280] Apparently this north–south discontinuity in North America finds its parallel in eastern Asia also, where scattered outposts of temperate zone plants occur in tropical latitudes.[256]

Considering the previous disruption in temperate latitudes that resulted from the Oligocene development of aridity east of the Rockies, the temperate flora was thus fragmented into three areas (Table 4). Table 4 emphasizes only floristic features common to all three segregates. It does not reveal the fact that Middle America has closer ties with southeastern North America than with west-central North America. This tripartite distribution is also reflected in bryophytes and in animals such as Vaux's swift, barred owl, spotted owl, and vicarious subspecies of the scrub jay and pigmy nuthatch.

Fig. 99.   Lush growth of vascular epiphytes in the foggy interior of a forest dominated by *Liquidambar styraciflua mexicana* and *Pinus strobus chiapensis* in Mexico.

The bulk of the temperate zone vasculares in Middle America occur as an enrichment of what will later be treated as midmontane rainforest. For example, in part of Chiapas which was collected rather thoroughly, 14.5% of the species and 49.6% of the genera were shared with southeastern North America.[64] This is the elevational range which coincides with the vertical thickness of the tropical cloud stratum, and where cloud makes contact with mountain sides there is much fogginess, so the name "fog forest" is appropriate. The temperate zone trees that occur in this belt tend to dominate the tallest stratum, with shorter evergreen trees such as *Podocarpus*, as well as the ground flora, being mainly of tropical affinity. Thus, tree ferns and an abundance of vascular epiphytes occur where the overstory is dominated by *Liquidambar styraciflua mexicana* and *Pinus strobus chiapensis* (Fig. 99).

THE TEMPERATE XEROPHYTIC FOREST REGION

Communities dominated by xerophytic trees of the Madro–Tertiary Geoflora came to dominate areas located between steppe or chaparral and temperate mesophytic forest in western North America. Those areas in which such forests occur as climatic climaxes will be treated as the Temperate Xerophytic Forest Region. Three Provinces occur in the foothills of the Cordillera.

The *Pinus ponderosa* Province

The *Pinus ponderosa* Province is distinguished by forests in which one or more of the following closely related taxa are climax dominants: *Pinus durangensis, P. engelmannii, P. leiophylla,* and *P. ponderosa* (Fig. 100). It is a common observation that plants which are climax under a given moisture regime play the role of seral opportunists in contiguous wetter environments where they eventually succumb to superior competitors. In conformity with this synecologic principle, the trees listed above are seral in secondary succession in the contiguous *Pseudotsuga menziesii* Province where fires have been frequent, and for this the distinctiveness of the *Pinus ponderosa* Province has usually been overlooked. At its northern extremity the Province reaches almost to central British Columbia. On the south it extends well into Mexico.

In a Section west of the Sierras and southern Cascades the *Pinus ponderosa* forests commonly have a lower tree layer composed of the deciduous *Quercus kelloggii* or the evergreen *Q. chrysolepis*. There the undergrowth is rich in *Arctostaphylos, Ceanothus,* and other evergreen shrubs which are dominants in the chaparral vegetation in slightly drier climates. Such shrubs are quick to multiply following fire or logging, and with repeated disturbance they thicken until natural tree invasion is extremely slow.

Fig. 100. Climax *Pinus ponderosa–Arctostaphylos uva-ursi* forest at the foot of Laramie Peak, Wyoming.

East of the Sierra–Cascade axis the Province is represented on most but not all ranges of the Rocky Mountains. In a Section extending from central Arizona northward a lower tree layer is lacking and *Pinus ponderosa* is the sole arborescent species. The undergrowth here is variously composed of shrubs (*Arctostaphylos uva-ursi, Ceanothus, Purshia tridentata, Quercus, Symphoricarpos*) in certain habitat types, and of grasses (*Agropyron, Andropogon, Bouteloua, Festuca, Muhlenbergia, Poa*) in others.

From central Arizona southward any of the pines listed earlier may participate in the dominant stratum, but usually not in the same area. *Quercus* or *Arbutus* commonly provide a lower tree stratum.

The forests in this Province occur on Prairie, Western Brown Forest, and Gray–Brown Podzolic soils.

For several months during summer (Figs. 12CD) the forest floor in this Province is dry enough to burn readily, and especially where grasses form the ground cover, slow-spreading surface fires are common. Recurrent fires kill most undergrowth shrubs and herbs back only to the ground surface. Nor do they harm old trees owing to their thick, fire-resistant bark. But by reducing the numbers of seedlings that can attain sufficient size to endure these fires, burning keeps the stands open.[416] In pre-Caucasian time natural fires together with those started by the aborigines kept the stands in a more open, savannalike condition than is common now.

On moist terraces of swales in this Province *Populus tremuloides* forms climax stands rich in mesophytic herbs or shrubs. Larger species in this

Fig. 101.   Interfingering of vegetation belts from alpine tundra to steppe. Villa Grove, Colorado.

genus fringe the streambanks. Alluvial soils in narrow, protected ravines may support more complex deciduous forests composed variously of *Acer negundo, Alnus oblongifolia,* and *Crataegus,* along with *Populus.* Other nonzonal soils result in parks, often of large sizes, in which either grasses or species of *Artemisia* are the dominants.

The forests of this Province occur between those of the *Pseudotsuga* Province on the wetter side, and steppe, chaparral, encinal, or the *Pinus cembroides* Province on the drier side. Wherever xerophytic forest is bounded on the dry margin by steppe, the outposts are restricted to rock outcrops, stony soils, or sands. At other dry ecotones xerophytic forest becomes restricted to ravines or north-facing slopes (Fig. 101).

Timber production and livestock grazing are the principal land-uses in this Province.

The *Pinus cembroides* Province[3]

In an area bounded by the Mexican State of Hidalgo, eastern California, central Oregon, southern Alberta, and western Texas there occur woodlands or savannas dominated by various combinations of pines in the subsection *Cembroides* (*Pinus cembroides, P. edulis, P. monophylla, P. quadrifolia*), by *P. flexilis* (low-altitude ecotypes), or by species of *Juniperus* (*J. californica, J. deppeana, J. flaccida, J. monosperma, J. occidentalis, J. osteosperma, J. scopulorum,* etc.), all of which have a significant degree of ecologic equivalence, judging from their similar positions in vegetation catenas. *Juniperus* is the more prevalent genus in the lower or drier margin, with pines becoming dominant higher in the foothills. From Texas to southern California the communities may be locally enriched by populations of *Cupressus arizonica*.

Only in a considerably smaller area around the southern Rockies do these woodlands occur on zonal soils (Sierozem, Brown, Chestnut, Western Brown Forest). Since the appropriate climate (Fig. 12E) occurs around the base of each of the many small mountain ranges, the type is highly discontinuous even here, i.e., in New Mexico, northern Arizona, and southern Utah.

The trees are of low stature, usually under 7 m tall, rarely to 20 m, and the *Juniperus* is commonly multiple stemmed at low elevations in the Province (Fig. 102). Although they may form an essentially closed canopy that practically excludes undergrowth, usually they are widely spaced as in a savanna. In the savannas the continuous phase is floristically similar to that of the contiguous steppe or chaparral in slightly drier environments, with the communities representing a superposition of trees over vegetation layers which alone make up steppe or chaparral. The interstitial plants are therefore mainly perennial grasses in *Agropyron, Bouteloua, Festuca, Hilaria, Muhlenbergia, Poa,* or *Stipa,* or shrubs such as *Arctostaphylos, Artemisia, Cercocarpus, Purshia, Quercus,* or *Rhus.*

Two divisions may be recognized. From southern Idaho to central Wyoming and northward to southwestern Alberta, the only pine associated with or replacing *Juniperus* is *Pinus flexilis* (Fig. 103). Throughout this division the

Fig. 102. *Pinus edulis–Juniperus monosperma* savanna near Rowe, New Mexico. The interstitial herbaceous vegetation, dominated by *Bouteloua gracilis*, has been shortened by heavy grazing.

Fig. 103. Thin stony soil supporting *Pinus flexilis–Juniperus scopulorum* savanna, with *Artemisia tridentata vaseyana* dominating the vegetation on less stony soil. Near Choteau, Montana.

stands occur on stony soils as edaphic climaxes. From southern Wyoming southward all pines belong to the Cembroides subsection.

The warm-dry to cold-wet sequence of Provinces characteristic of the Rocky Mountains (*Pinus cembroides, P. ponderosa, Pseudotsuga, Picea engelmannii* Provinces) is frequently truncated from below.[98] Thus, the *Pinus cembroides* belt is often missing, commonly the *P. ponderosa* belt too is missing, and occasionally even the *Pseudotsuga* belt is lacking, so that in a few places steppe extends from the basal plain to high altitudes where it gives way to forests of the *Picea engelmannii* Province.

Where woodland or savanna occurs on small mountains rising from extensive steppe, it may be confined to protected slopes, or on flatter topography to rock outcroppings or to sands. Notable in these low-altitude fragments is the erratic absence of species expected in such communities, this probably being a consequence of erratic patterns of Hypsithermal extinctions as species were "squeezed off" the hilltops, coupled with varying abilities of the plants to subsequently migrate across inhospitable stretches of steppe and regain lost ground.

All trees in this vegetation are easily killed by fire, but the ground vegetation is usually so thin, especially where vegetation of this type occurs as edaphic climaxes on stony soils, that fire is seldom devastating. *Juniperus* is widely disseminated by birds, and in many places it has been reported as invading grassland. These situations may represent recolonization of burned areas which had a rather heavy ground cover,[212] or a recovery of territory lost during Hypsithermal drought, or may represent a shift in ecotones resulting from livestock destruction of grass which upset competitive balances. Where the woodland or savanna abuts chaparral, repeated burning favors a thickening of the chaparral shrubs which had formed the interstitial cover.

Cembroides pines furnish large edible seeds of some commercial value, and the young trees are commonly used for Christmas decoration. *Juniperus* wood is highly rot-resistant, and has been widely used for fence posts. The foliage and berries of this tree are heavily used by deer, and the berries are used by birds. Much use was formerly made of both genera of trees for firewood.

The grasses and some of the shrubs in this vegetation are important as forage for livestock, and since their productivity is increased at least for a time after eliminating the conifers, considerable interest is being directed to kill all the trees by mechanical means or fire.[74] This is of course devastating to many forms of vertebrate and invertebrate life that depend upon the trees, and inevitably increases soil losses through erosion by wind and water. Long-term effects are unknown at present.

Owing to the savannalike physiognomy of so much of the Temperate Xerophytic Forest, and especially the *Pinus cembroides* Province, the characteristic vertebrate inhabitants differ somewhat from those of the coniferous forests of cooler and wetter areas. Some representatives of this fauna are wapiti, bison, mule deer, bighorn sheep, porcupine, cottontail rabbit, woodrat, squirrel, puma, bobcat, badger, weasel, hawk, magpie, flicker, mountain bluebird, gray titmouse, pinon jay, horned lizard, collared lizard, and brown-shouldered lizard.

The Encinal Province

At many places in the foothills of western North America, chiefly where the winters are rather severely cold, the *Pinus cembroides* Province occurs on the drier side of the *Pinus ponderosa* Province. But in other places, especially where the winters are clearly more mild, savanna or woodland dominated by various species of *Quercus*, i.e., encinal,* occurs in place of the *Pinus cembroides* belt. This substitution occurs from southwestern Oregon to southern California, and from southeastern Arizona and western Texas to central Mexico.

The earliest suggestion of encinal is a fossil flora of Eocene age from northeastern Nevada, which contains *Mahonia*, *Pinus*, sclerophyllous *Quercus*, and *Vauquelinia*. When the Sierras uplifted, the Great Basin became

---

*The Spanish word *encino* means "oak," and the suffix *al* means "area of."

too cold for this type of vegetation and it receded southward, although a large portion of it persisted to the windward of the Sierras.

Some of the encinal trees are deciduous and some evergreen. All are of low stature. The diversity of the genus *Quercus* is highest in Mexico where 112 species have been recognized. Usually conifers are lacking, but in California three are minor members: *Pinus coulteri, P. sabiniana,* and *Pseudotsuga macrocarpa.*

Homologous vegetation around the Mediterranean Sea includes dominants in *Arbutus, Ceratonia, Olea, Pinus halpensis, Pistachia,* and *Quercus,* with the last tending to dominate. Much of the woodland and savanna of that area has been reduced to unpalatable scrub by centuries of land abuse, especially devastation by goat browsing.[131]

The Siskiyou Section of the encinal Province is very small and confined to the dry valleys of the Siskiyou Mountains of southern Oregon and adjacent California. The arborescent stratum there consists almost entirely of *Quercus garryana* and *Q. kelloggii,* both deciduous, and the evergreen *Arbutus menziesii. Acer macrophyllum* and *Umbellularia californica* occur as riparian trees.

The extensive Californian Section occurs in the foothills surrounding the great interior valley of that state, and on both sides of the southern Coast Ranges. The tree stratum is variously composed of the deciduous *Aesculus californica, Quercus douglasii, Q. garryana, Q. kelloggii,* and *Q. lobata,* and the evergreen *Arbutus menziesii, Quercus engelmannii, Q. wislizenii, Lithocarpus densiflorus, Pinus coulteri, P. sabiniana,* and *Pseudotsuga macrocarpa.*[319]

On gentle slopes low in the foothills encinal soils are typically Noncalcic Brown, and the trees occur in savanna spacing. Perennial grasses of much the same species as in nearby steppe once formed the ground cover, but as in the steppe these have been replaced by alien annuals (Fig. 104). On steeper and more stony soils there are edaphic climaxes in which the trees grow close together as in woodland, with shrubs of chaparral character prevailing in the undergrowth (Fig. 105). A heavy layer of litter occurs here, since the evergreen leaves are sclerophyllous.

The trees show little tendency to sprout after a fire, and where the soil is stony fire stimulates the sprouting and germination of associated shrubs, so that repeated burning tends to convert the woodland into a fire climax composed of species primarily characteristic of continguous chaparral (Fig. 106). On deeper soils where there is a grass understory, repeated burning tends to convert savanna to a fire climax of grassland with a few fire-tolerant shrubs. Burning is now practiced to divert all productivity possible into annual grasses.

Riparian trees in the California Section include *Acer macrophyllum, Aesculus californica, Alnus rhombifolia, Platanus racemosa, Populus fremontii,* and *Umbellularia californica.*

Fig. 104. *Quercus douglasii–Pinus sabiniana* (the taller tree) savanna near Redding, California.

Olives, grapes, almonds, and cork oak are plants adapted to encinal climates in this Section, but only the first two are grown extensively. Otherwise the vegetation is used mainly for beef production or deer browse.

East of the desert that lies astride the California–Arizona border is another, the interior, Section of the encinal Province. From southeastern Arizona and western Texas it extends southward far into Mexico. North of

Fig. 105. Interior of woodland with *Quercus wislizenii* and *Arbutus menziesii* forming the overstory, and *Rhus triloba* and *R. diversiloba* beneath. Santa Lucia Mountains, California.

Fig. 106. Three-year-old sprouts of *Adenostoma fasciculatum* (unique among chaparral shrubs for its small linear leaves) in a habitat formerly supporting woodland. Merced, California.

the international border the principal trees, all evergreen, are *Arbutus arizonica, A. texana, Quercus emoryi, Q. grisea, Q. oblongifolia,* and *Vauquelinia californica.* As in California, both savanna (Fig. 107) and woodland are represented, but unlike the California Section, alien annuals have not assumed dominance in these savannas, and perennial species in *Andropogon, Bouteloua, Eragrostis,* etc. still dominate the herb stratum even though

Fig. 107. Savanna dominated by *Quercus emoryi*, with *Bouteloua* and *Andropogon* prominent in the herbaceous layer. West of Nogales, Arizona.

the major use of the vegetation has been grazing. Streamside trees here in the interior include *Fraxinus velutina, Juglans rupestris, Platanus wrightii, Populus fremontii,* and *Salix gooddingii.*

Aside from producing good grazing for cattle, this Section provides habitat especially favorable for deer, peccary, turkey, and wild pigeon.

In all three Sections of the encinal Province species of *Quercus* were well used by the aborigines who had only to crush the fruits and leach out excessive tannin to obtain an abundance of starch.

In the southwestern United States this Province interfingers with steppe or chaparral at lower elevations (Fig. 108), and with the *Pinus ponderosa* Province above. In the rain shadow of the Sierra Madre Occidentale in Sinaloa, *Quercus* savanna covers the mountain slopes above tropical xerophytic forest.[153] On the west slope of the Sierra Madre Oriental in Tamaulipas *Quercus-Pinus* savanna occurs just above the desertic basal plain.[277]

Along the dry margin of the *Acer saccharum* and *Quercus falcata* Provinces of eastern North America, the savannas of *Quercus macrocarpa* or *Q. marilandica–Q. stellata* occupy positions ecologically equivalent to the temperate xerophytic forest of the Cordillera, and are especially similar in physiognomy to encinal. However, these wholly deciduous communities are considered ecotonal to the temperate mesophytic forest since their dominants without exception represent a segregation of species widely distributed over the mesophytic forests and are neither taxonomically nor historically related to the physiognomically identical forests of encinal.

Fig. 108. Encinal gives way to chaparral on relatively dry ridges and slopes on the east side of the Santa Lucia Mountains, California.

THE CHAPARRAL REGION[295,379]

Chaparral is that vegetation of extratropical latitudes dominated by shrubs mostly 0.5–3.0 m tall that are mostly evergreen and sclerophyllous, occupying areas which are only slightly less dry than steppe or desert. Chaparro is a Spanish word meaning scrubby evergreen *Quercus,* but chaparral will be used here for a variety of vegetation of this character, whether dominated by *Quercus* or not.

For reasons never made quite clear, in describing Californian vegetation North American authors commonly include the encinal of contiguous less arid environments with chaparral. Some confusion may have arisen since the shrub layer that dominates chaparral often occurs as an inferior layer in adjacent xerophytic woodland, but this is a common phenomenon by no means peculiar to chaparral ecotones.

Some Sections of chaparral are second only to tropical rainforest in their species diversity.[314] Most of these shrubs have evergreen leaves larger than nanophylls, and in North America leaf margins are frequently spiny (Fig. 109). Otherwise spinescence, succulence, and bud scales are rarely encountered. Etherial oils are common among the plants. Legumes are poorly represented.

Fig. 109. *Quercus dumosa* showing biotype variation. Spiny-margined leaves as at the right are common among different genera of chaparral shrubs in California. The scale is 30 mm long.

Origin and Distribution

All continents have this Region extending inland from their west coasts along the poleward fringe of the subtropical dry belt: in southwestern North America (where it is called chaparral), in northern Chile (mattaral), in southwestern Australia (mallee scrub), at the Cape of Good Hope (fynbosch, karroo), and in the eastern Mediterranean area (garigue, macchia, maquis, phrygana, shibliak, tomillares).

North American and Mediterranean chaparral share several genera including *Artemisia, Quercus, Rhamnus,* and *Rhus,* but only two genera are shared with Chile. The Australian and African areas share the *Proteaceae,* and the Chilean and African areas share a few genera. These limited sharings reflect neither a common origin nor floristic exchanges. Apparently each of the widely separated units has differentiated out of local taxonomic stocks, which in a few cases happened to bear resemblances. Since in the strict sense the word chaparral refers to *Quercus,* a purist has a theoretical basis for objecting to the use of the term in a broader sense to include the mallee scrub dominated by *Eucalyptus* in Australia. However, any term he might offer to cover all the geographic units would have to be so ambiguous as to include vegetation that is very different in its ecology!

North American chaparral evolved mainly from the Madro–Tertiary Geoflora. Its taxonomic affinities are, therefore, more with the Neo–Tropical Tertiary rather than the Arcto–Tertiary Geoflora.

By late Miocene time chaparral had clearly differentiated, and was continuous from the Pacific coast to Arizona.[16] Early in Pliocene time aridity to the lee of the slightly elevated Cascades provided suitable climate for this vegetation in the Great Basin area, but then with further uplift of those mountains the winters became more severe and this brought about a drastic purge of the Great Basin chaparral. Only a strongly impoverished remnant has persisted in the foothills between the Sierra–Cascade axis and the Rockies, but from New Mexico to the Pacific coast chaparral has maintained moderate to high species diversity.

Environment

West Coast areas supporting chaparral are situated in the zone of transition between the Westerlies and the Subtropical High Pressure belt. Since these belts shift following the sun, winter brings the Westerlies closer to the Equator so that onshore winds provide precipitation in the range of 300–900 mm during the cool season, and temperatures remain mostly at levels permitting photosynthesis. A reverse migration of these climatic belts in summer brings chaparral under the influence of the expanded Subtropical High Pressure belt, so there is a rainless period during at least the early part of summer (Fig. 12G–J), and plant functions are curtailed after a brief period of spring flowering. Climates of this character which have mild winters are popularly referred to as Mediterranean climates, but chaparral is by no means restricted to climates with such mild winters. Furthermore, the same climatic type with mild winters characterizes parts of the steppe, encinal, and other types of forest.

Along coasts chaparral gives way to desert in an equatorial direction, as a result of the lengthening of the dry season. In a poleward direction the severity of summer drouth diminishes until encinal takes over the landscape.

Inland extensions also have dry periods in summer, but they may have biseasonal rainfall as in Arizona (Fig. 12I) and Greece, or have rainfall rather uniformly distributed throughout the year as in southern Africa. In the interior parts winters can be quite cold (Fig. 12J), whereas along sea coasts winters are mild, although subject to frequent frost.

The *Quercus dumosa* Province[408]

The *Quercus dumosa* Province consists of a coastal Section centered on southwestern California (Fig. 110) and an inland Section centered on southern Arizona (Fig. 111), with a broad tract of desert intervening.

In the coastal Section the following shrubs are abundant: *Adenostoma fasciculatum, Arctostaphylos, Ceanothus, Cercocarpus betuloides* (semideciduous), *Fremontia californica, Photinia arbutifolia, Pickeringia montana, Prunus ilicifolia, Quercus dumosa* (s.1), *Rhamnus,* and *Rhus*

Fig. 110. Chaparral with denuded firebreaks, near San Bernardino, California.

Fig. 111.   Chaparral dominated by *Quercus* and *Arctostaphylos*, near Bisbee, Arizona.

ovata. These are deep-rooted shrubs,[188] averaging about 2 m in height. Since they are hard-wooded, sclerophyllous, and their canopies often cover the ground completely, they usually present formidable interference for the traveler on foot. The shade they cast allows only a sparse understory of mainly annuals, although in African, Australian, and Mediterranean areas geophytes are reported as abundant. Shrub communities of this general character prevail on Noncalcic Brown Soils over a large area in southwestern California, extending southward into Mexico and northward along the foot-hills of the Sacramento–San Joaquin Valley.

The firm-textured leaves of chaparral shrubs resist decay, so litter accumulates to depths up to 15 cm, and this helps make the vegetation highly flammable during the hot dry summers.[231] Pronounced adaptations to fire bear mute evidences that this has long been an important environmental factor. Most of the woody plants are killed back only to the ground by fire, and as a result of repeated postfire regeneration, the summit of the tap root enlarges irregularly to form a lignotuber beset with adventitious buds that send up new shoots promptly after each fire. Many species, whether capable of sprouting or not, have seeds that lie dormant in the soil surface for many years until stimulated to germinate by changes in their structure or physiology that are induced by the heat of vegetation fires.[201]

Herbaceous plants, especially annual grasses, dominate for a few years after a fire. By then the woody plants have regained their position of eminence by vigorous sprouting, or by new germinations,[350] and the herbaceous element has become very attenuated. About half the species are able to sprout, and these, especially *Adenostoma*, increase their proportionate representation quickly. Perhaps the abundance of *Adenostoma* at low altitudes in the chaparral belt is largely a reflection of more frequent fires there. The other shrubs, regenerating by seedlings, slowly regain their relative positions as the years pass. The normal physiognomy of the community is restored in 8–10 years, and species readjustments are essentially complete by about 30 years.[172] Then, should an area escape fire for 50 years the vegetation becomes somewhat decadent and productivity declines, although there is no evidence of replacement by another community. An immobilization of nutrients in organic residues, along with accumulations of toxic products of some of the species[274] seem to be involved here, for even as the old community thins new seedlings do not get abundantly established until a new fire sweeps the area. It has been opined that evolution may have favored flammability of chaparral shrubs, since germinating and sprouting are favored, as well as rejuvenation of old plants.

In the main body of the coastal Section *Adenostoma* maintains a position of dominance in almost pure stands on poor soils.

The trees which line the streams include *Acer macrophyllum*, *Platanus racemosa*, *Populus fremontii*, and *Umbellularia californica*. On protected slopes and in deep valleys scrub gives way to encinal. As edaphic climaxes on stony soils, or topographic climaxes on exposed slopes, chaparral communities with similar structure and composition extend as far north as the Rogue and Umpqua Valleys in southern Oregon.[114]

This coastal Section of the *Quercus dumosa* Province forms ecotones with the *Artemisia californica* Province, steppe, desert, and all three of the Temperate Xerophytic Forest Provinces. In many places it is evident that fire has allowed chaparral dominants to replace woodland or forest in the contiguous foothills (Fig. 106), for the dominants of those Provinces do not sprout, and a few successive fires can eliminate seeds sources with the thickened shrubbery then presenting formidable competition that retards tree invasion.

Before extensive urbanization usurped so much of the irrigable zonal soils, citrus fruits were a major agricultural product of this Section. The nonirrigable land that remains is now of value mainly for watershed protection and deer browse. Attempts to eradicate chaparral in hilly country to create grasslands for cattle grazing and to increase water yield,[30] have aggrevated erosion[237] and in places magnified a natural tendency for landslides to occur.[21,337] Where chaparral is climax there may be some justification for burning at intervals of about 15–20 years to prevent excessive accumulations of fuel that could support devastating holocausts.

The inland Section of the *Quercus dumosa* Province is closely similar to the coastal Section in physiognomy (Fig. 111), ecology,[322] and taxonomy.[252,305] Shrubs shared at the species level are

Arctostaphylos pungens	*Quercus dumosa*
Ceanothus greggii	*Q. palmeri*
C. intergerrimus	*Rhamnus californica*
Eriodictyon angustifolium	*R. crocea*
Eriogonum wrightii	*Rhus ovata*
Fremontia californica	*R. trilobata* (deciduous)

Other common shrubs of the interior include:

Arctostaphylos pringlei	*Cowania stansburiana*
Berberis fremontii	*Garrya flavescens*
B. haematocarpa	*G. wrightii*
Cercocarpus breviflorus	*Q. grisea*
C. montanus	*Q. undulata*

The foregoing species, nearly all evergreen, represent chaparral as it occurs in Arizona, with taxonomic diversity and areal representation diminishing eastward to western Texas. To the south, in the rain shadow of the Sierra Madre Oriental many of the same genera but few of the same species provide the dominants of a chaparral belt that extends up dry ridges into the encinal.[277,300,301]

As on the coast, owing to prompt sprouting or stimulated germination, chaparral shrubs in the inland Section regenerate quickly after fire, with herbaceous vegetation providing temporary cover.[315]

This Section forms ecotones with desert or the *Bouteloua gracilis* Province of steppe on the drier margin, and with temperate xerophytic forest Provinces on the wetter margin.

Land use has been largely limited to deer management and livestock grazing, but relatively few of the shrubs provide browse except for deer. As in the coastal Section, there has been some interest here in destroying the shrubs with fire or chemicals in order to favor sown forage grasses and to increase water yield.[23] However, most chaparral stands occur on soil that is more stony than is favorable to the maintenance of grass in dry climates, and erosion appears to be increased significantly by such conversion.

The *Artemisia californica* Province

A small but distinctive type of vegetation that is related to chaparral by virtue of its climatic pattern and shrub dominance occurs in a narrow belt along the Pacific Ocean from central California to approximately Rosario in Baja, California.[129] This differs from the adjacent *Quercus dumosa* Province on its landward side by having lower rainfall (Fig. 12H) and a distinctive flora of shrubs that are shorter, mostly summer-deciduous and fire sensitive.[172] Associates of the most characteristic species, *Artemisia californica*, include *Encelia californica, Eriodictyon, Eriogonum fasciculatum, Eriophyllum confertifolium, Haplopappus squarrosus, Horkelia, Lotus, Rhus integrifolia* (evergreen), *Salvia*, and *Viguiera laciniata*.

Most of the plants are no more than 1 m tall (Fig. 112), soft-wooded or only half-shrubby, and the foliage is grayish rather than dark or olive green as in the *Quercus dumosa* Province. Also these plants are more microphyllous, more pubescent, they do not cover the ground so completely, and they are accompanied by many perennial and annual herbs.

Just southeast of Los Angeles two altitudinal belts have been distinguished in this Province.[319] The lower, dominated by *Artemisia californica* and

Fig. 112.   Vegetation dominated by *Artemisia californicum*, near Laguna Beach, California.

Eriogonum fasciculatum, has a well-developed understory mainly of annual grasses. The upper, dominated by *Eriogonum fasciculatum* and *Rhus laurina*, has a negligible grass component.

In California this Province extends from sea level to as much as 600 m above before giving way to vegetation of the *Quercus dumosa* Province. At its southern extremity in Baja California, desert intervenes between this vegetation and the sea with its elevational range lifted to between approximately 600–1500 m.[180] Disjunct extensions of this type of chaparral occur inland in the latitude of Monterrey and Los Angeles, where outpost communities occur on excessively drained gravel in the rain shadow of the Coast Range up to 80 km from the ocean.

Most of the shrubs of this Province temporarily invade habitats belonging to the *Quercus dumosa* Province following fire.

The *Quercus gambelii* Province

The *Quercus gambelii* Province occurs in the foothills of the Great Basin area, and on the eastern margin of the Rocky Mountains at the same latitude. Nearly all the dominants in this Province also occur in the *Quercus dumosa* Province, so it is clearly an attenuation of the latter, consisting of the most cold-tolerant taxa (compare Figs. 12I,J). During Hypsithermal time *Quercus gambelii* extended its range northward along the foothills east of the Rockies as far as the Black Hills of South Dakota, where it hybridized with *Q. macrocarpa* relics from Wisconsin time. Later, it and probably other thermophilic associates, receded southward, leaving a stranded hybrid swarm as sole evidence of the temporary northward range extension.[285] West of the Rockies *Quercus dumosa turbinella* likewise extended northward during the

Hypsithermal Interval, leaving hybrids with *Q. gambelii* as evidence.[82] In the same way hybrids of *Cowania stansburiana* × *Purshia tridentata* have been found in the Great Basin well to the north of the present range of the former species.

Where the Province is well represented west of the Rockies *Quercus gambelii* is the outstanding dominant. Toward the lower altitudinal limit of the belt this tall deciduous shrub occurs in small clonal groves which in the aggregate cover about half the surface of gentle slopes (Fig. 113). *Carex geyeri*, *Symphoricarpos oreophilus*, and *Rosa* are characteristic of the low-growing matrix, but under heavy grazing *Poa pratensis* becomes dominant. On steeper ground at higher elevations *Quercus gambelii* is a member of a nearly closed community (Fig. 114) associated variously with other shrubs, some of which are deciduous (*Acer grandidentatum*, *Amelanchier alnifolia*, *Cercocarpus montanus*, *Prunus virginiana melanocarpa*, *Rhus trilobata*), and some evergreen (*Arctostaphylos patula*, *Ceanothus velutinus*, *Cercocarpus ledifolius*). Apparently in this cold Province the deciduous habit is of more advantage than it is in the *Quercus dumosa* Province to the south. *Cercocarpus ledifolius* and *Amelanchier alnifolia* extend north of the Great Basin as far as southern Washington and central Idaho, forming relatively pure stands on thin stony soils in the vicinity of lower timberline.

On the east flank of the Rockies chaparral is well represented as far north as central Colorado by a relatively rich flora of *Cercocarpus montanus*, *Holodiscus dumosus*, *Prunus virginiana melanocarpa*, *Quercus fendleri*, *Q. gambelii*, *Q. gunnisoni*, *Rhus trilobata*, and *Rubus deliciosus* (Fig. 115). As

Fig. 113.   Clones of *Quercus gambelii*, near Springville, Utah.

Fig. 114. *Quercus gambelii* and *Acer grandidentatum* forming chaparral, near Salt Lake City, Utah.

Fig. 115. Chaparral dominated by *Quercus* along the east flank of the Rocky Mountains, near Castle Rock, Colorado.

edaphic climaxes chaparral vegetation extends as far north as Montana, with *Cercocarpus montanus* or *Rhus trilobata* forming thickets on thin soils in the vicinity of lower timberline (Fig. 116). All species east of the Rockies are deciduous.

Streams that descend through chaparral vegetation are bordered by deciduous trees and shrubs that include *Alnus tenuifolia, Acer negundo, Betula fontinalis, Cornus stolonifera, Crataegus, Physocarpus malvaceus, Populus angustifolia,* and *Salix.*

Except for *Quercus,* which is usually the outstanding dominant, most of the shrubs provide useful browse. Where the shrubs are clustered in groves grass and other herbs in the interstices can provide up to 3.6 kg/ha of forage.[53]

At its upper altitudinal limits chaparral abuts forests of the *Pinus ponderosa* Province, or where such is lacking, the *Pseudotsuga* Province. Below, soils permitting, it gives way to steppe everywhere. A curious feature of the vegetation pattern in this area is that on those slopes where the *Quercus gambelii* Province is represented, the *Pinus cembroides* Province is usually lacking, and vice versa. There is some evidence that this alternation is climatically determined, with chaparral prevailing where there is a higher proportion of cool-season precipitation.[71]

THE STEPPE REGION ✳

Steppe is the extratropical grassland of areas where the zonal soils are too dry for trees, and herbaceous perennial grasses are well represented. It is only one of a number of types of the physiognomic category of grassland.

The dominant grasses of steppe vary greatly in height, but all die back to the ground each year. They may be rhizomatous so that a continuous or interrupted sod is formed, or they may be caespitose, forming "bunchgrass" or "tussock" grassland. Forbs are of minor importance in the drier parts of steppe, but toward the wetter edge they become conspicuous, and may even exceed the graminoids in dry-matter production. Such forb-rich steppe is called meadow steppe. When present, shrubs may be dwarfed, shorter than the herbs and interspersed among them; they may be aggregated into thickets confined to relatively moist microenvironments; or they may rise above the grasses and form a discontinuous upper layer—a physiognomic type referred to as shrub steppe. Approximately 50% of the steppe flora consists of hemicryptophytes, with chamaephytes and therophytes each contributing about 20% of the species.

Broad strips of gallery forest commonly occupy the subirrigated soils of riparian habitats in the less-arid margin of a steppe area, but trees dwindle to narrow and discontinuous streamside fringes of limited taxonomic variety in more arid parts.

In North American steppe about 100–400 gm dry matter are produced per square meter per year,[101,389] only half of which can be harvested by grazing

Fig. 116. *Cercocarpus montanus* and *Rhus trilobata* forming an edaphic climax on thin soils in steppe where *Stipa comata* and *Bouteloua gracilis* dominate zonal soils. Scattered *Pinus ponderosa* and *Juniperus scopulorum* occur in ravines on the distant ridge. Near Newcastle, Wyoming.

animals without damaging the vegetation. Excessive grazing pressure has led to the widespread replacement of the native herbs of high forage value with other plants, native or alien, that are either nonacceptable to grazing animals, or are not very available owing to low stature or short seasons of growth. Annuals nearly always become conspicuous, but shrubs and perennial forbs too may figure prominently in this secondary vegetation.

Many terms have been used for vegetation with grassland physiognomy in different parts of the world. In North America the word "prairie" has been applied to fire-maintained grassland, meadow-steppe, shrub steppe, scrub, salt marsh, tundra, and groveland. In the tropics all grasslands, free of woody plants or not, are generally called "savanna." South African grassland and savanna have traditionally been called "veld." Locally in South America the terms "campo limpo" and "pampa" have been applied. Europeans have long used the term steppe for climatically determined grasslands, applying the term "meadow" to the mesophytic grasslands on special soils in forest Regions, and "grass heath" if such enclaves are relatively xerophytic. Fen and marsh are still other types of grassland that differ from steppe ecologically.

Origin and Distribution

The most significant evolutionary advance in the angiosperms since Cretaceous time has been the development of the herbaceous life form, and especially the annual habit. Although fossil *Gramineae* have been reported from Eocene deposits, herbivores with dentition suggestive of grass-eaters, and hence of savanna vegetation, did not appear abundantly until Oligocene time. In North America large areas of steppe are not believed to have developed until Pliocene time,[341] when the Rockies approached their present height[117] and produced a rain shadow that eliminated forest from uplands to the east. During glacial periods low temperatures made rainfall much more effective so that steppe invaded desert along its dry margin while giving way to forest on the wet margin. In Europe it is believed that the last interglacial may have become so warm that steppe expanded beyond its present limits[147]—a phenomenon that certainly happened during the Hypsithermal Interval of the Holocene Epoch. During the Hypsithermal Interval desert must have extended northward displacing steppe in the most extreme part of the rain shadow along the east base of the Rockies.

Steppe is found on all continents and large islands in temperate latitudes, typically occupying a position between desert and either forest or chaparral. In the temperate zone dryness sufficient to exceed the tolerance limits of forest or chaparral usually occurs well inland from oceans, as in the central parts of North America and Eurasia. But steppe may occur near the ocean in the rain shadow of mountains, as in California and Argentina.

In North America the Steppe Region is represented by several Provinces (see Fig. 55) that differ either as to their climatic pattern, or have had distinctive histories insofar as their major dominants are concerned.

Environment

In the Steppe Region there is always a season when low rainfall coupled with high evapotranspiration desiccates the soil sufficient to kill all naturally germinating tree seedlings. Unlike trees, many herbs can aestivate when the soil becomes depleted of its biologically useful water.[102] While the herbs are in aestivation the vegetation is highly flammable, and fire has always been an important factor determining the composition of steppe vegetation. Where there were large populations of ungulates their grazing pressure too has been important in shaping the evolution of species and of communities.

Almost equal in importance to the limited quantity of precipitation is its variability—a parameter which increases in magnitude as the annual total decreases. Accordingly there is an oscillation in vegetation composition, with relatively mesophytic species gaining in relative dominance during a series of wet years, then yielding to their associates during drouths.[418]

In the wet margin of steppe, surface moisture becomes continuous with subsoil moisture during the rainy season, but subsequently drouth extends downward too rapidly to allow tree seedlings to survive. At the dry margin of steppe rains often fail to penetrate below the reach of grass roots, so the low subsoil moisture storage in consequence of erratic recharge prevents the survival of all but a few species of forbs and shrubs that have deep tap roots.

In consequence of the low precipitation, there is very little loss of plant nutrients by leaching. The calcium that is so abundantly released in early stages of the weathering of primary minerals is converted to soluble bicarbonate by carbonic acid liberated from plant roots and microbes, then moves downward to approximately the depth of moisture penetration where it reverts to monocarbonate and accumulates as either a diffuse (see Fig. 56) or a cemented layer called caliche. In progressing from meadow steppe toward desert this carbonate layer tends to become more firmly cemented and nearer the surface. So much carbonate keeps the soil reaction circumneutral or slightly basic, but there is little chloride or sulfate in the zonal soils since these highly soluble salts are moved below the solum in those years when exceptional precipitation wets the profile deeply.

Owing to the numerous short-lived roots of grasses, the soils supporting grass tend to become darkened with humus to a degree and depth that varies directly with precipitation. The high nutrient-retaining capacity of the soils resulting from their humus content is magnified by the influence of climate on mineral weathering, which in steppe environment results in a type of clay (montmorillonite) that has very high cation adsorbtion capacity. (The illite clay of temperate forest, subarctic forest, and tundra is definitely inferior in this respect.)

Granular soil structure, favored by abundant calcium and humus, renders steppe soils so porous that under pristine conditions there is no apparent erosion although there is obvious runoff during convectional storms.

In consequence of the long dry season when the shoots of the closed communities stand dead and dry, fire spreads rapidly over steppe. In general, fire does no more than consume shoots after their photosynthetic role has ended, and new shoots emerge as usual from the underground organs when the next rainy season commences. But in the most arid parts of the Steppe Region fire injures certain species so much that recovery may take several years if not decades.

Along the wet margin of the Steppe Region, fires that start in the grassland tend to erode the edge of contiguous forest, commonly producing a border of fire-induced grassland to the leeward of areas where climate is the primary cause of herb dominance (Fig. 55). Such derived grasslands are excluded from this account of climatic grasslands, and treated as fire climaxes in forest Regions discussed elsewhere. In North America, at least, there is abundant evidence of diverse types that distinguish areas where grassland has been induced by fire.

The uplift at the close of the Mesozoic Era that eliminated the midcontinental Cretaceous Sea of North America was not sufficient to create mountains high enough that a rain shadow affected vegetation patterns in early Cenozoic time. But in the Oligocene Epoch the Rocky Moutains began to rise sufficiently to have a very important effect. They intercepted the moisture-laden air carried inland off the Pacific Ocean by the Westerlies, and so created a vast rain shadow extending from Canada to Mexico. In the course of time these mountains also contributed a vast sheet of sediments which streams flowing eastward across the rain shadow deposited as the present Great Plains. These Plains descend gradually from a rather high elevation next to the mountains to the Mississippi lowlands to the east. They've come to support three of the steppe Provinces.

The *Andropogon scoparius* Province

The *Andropogon scoparius* Province occurs as a north–south belt occupying approximately the eastern half of the midcontinental steppe (Fig. 55). It extends from southern Manitoba to the central coastline of Texas,[227] and for a short distance on southward into Mexico along the gulf coast.[252]

The *Andropogoneae*, to which the characteristic grass dominants belong, is pantropical, with its highest species diversity centered on the equator. These and many other dominants of the *Andropogon* Province are widespread on sandy soils or other open habitats in the temperate mesophytic forest to the east. Thus in all probability the Province came into existence along the eastern edge of the Rockies as a preadapted residue, when the development of a rain shadow eliminated mesophytic trees from the uplands there. As aridity continued to intensify, forest withdrew even farther eastward and the *Andropogon* Province followed it as a belt, to be replaced by the still more xerophytic *Bouteloua gracilis* Province next the mountains.

Mt. + foothill meadows

During Wisconsin glaciation lowered temperatures increased the moisture effectivity so that the flora of the *Andropogon* Province spread westward to the Rockies once more, sharing the Great Plains with temperate forests species that also spread westward, with subarctic forest species that descended from the north,[56] and with Rocky Mountain forests that encroached from the west.[195,325] Then in Holocene time when temperatures rose once more to eliminate forest from the Great Plains and favor the reestablishment of the *Andropogon* Province in the eastern portion, relics of its previous western extension were left at the foothills of the Rockies.[290] Relics of the three forest floras were left scattered widely in the Great Plains. With the coming of the Hypsithermal Interval the eastward recession of the *Andropogon* Province continued, but chiefly along the main axis of the tract of the Westerlies, resulting in the "prairie peninsula" alluded to previously.

As is typical of continental interiors, in all three of the steppe Provinces east of the Rockies there is a warm-season maximum of convectional precipitation which centers on June (Fig. 12K).

A northern Section of the *Andropogon* Province may be recognized extending from the southern edge of Manitoba to central Texas. The most consistent dominant of the climatic climaxes is *Andropogon scoparius*, which forms an interrupted sod in association with *A. gerardi, Panicum virgatum,* and *Sorghastrum nutans.* On convex topography (Fig. 117) *Bouteloua curtipendula, Koeleria cristata, Sporobolus heterolepis,* and *Stipa spartea* are also well represented, whereas in depressions *Elymus canadensis* is favored, with *Andropogon gerardi Panicum virgatum,* and *Sorghastrum nutans* increasing at the expense of *Andropogon scoparius.* Large forbs such as *Amorpha canescens, Aster, Echinacea angustifolia, Erigeron, Helianthus, Hypoxis, Liatris, Petalostemon, Psoralea tenuiflora, Ratibida, Silphium,* and *Solidago* are well represented, giving this clearly the character of meadow–steppe. This type of grassland is so dense and tall that it was extensively mowed for hay before the bulk of the land was claimed by agriculture.

In contrast with the fire-maintained *Andropogon* vegetation of the "prairie peninsula" to the east, the soils here have a lime layer and are classed as Chernozem or Chestnut.

In the eastern part of the Section grazing or mowing favors the ascendency of *Poa pratensis,* but in the drier western parts *Bouteloua gracilis* and *Buchloe dactyloides* are the characteristic increasers.[4] Spring burning is commonly practiced for its detimental effect on the early-developing *Poa,* and its stimulating effect on the taller native grasses that start growth later and are more productive.

On moist slopes *Andropogon gerardi* assumes rather complete dominance, with its foliage growing up to a meter in height and the inflorescence commonly rising over 3 m. Owing to its superior growth in relatively moist environments, this is the species which dominated the Prairie Podzolic Soils

Fig. 117. Steppe dominated by *Stipa spartea* and *Andropogon,* with *Symphoricarpos* thickets. Near Rugby, North Dakota.

Fig. 118.   *Spartina pectinata* in the edge of a temporary pond, with *Andropogon gerardi* and *A. scoparius* on upland behind. Near Bristol, South Dakota.

of the "prairie peninsula," where *A. scoparius* was restricted mainly to sandier soils, to shallow soils atop cliffs, etc.

Marshes with little or no salinity support stands of *Spartina pectinata* (Fig. 118), with foliage 1–2 m tall, or shorter communities of *Carex, Eleocharis,* or *Muhlenbergia. Elymus canadensis* commonly forms a belt along the margins of such wet areas. In the few seasonal saline marshes may be found *Atriplex, Distichlis stricta, Hordeum jubatum, Puccinellia, Salicornia rubra, Scirpus paludosus,* and *Suaeda.*

Gallery forests of *Acer negundo, Celtis occidentalis, Fraxinus, Juglans, Ostrya virginiana, Populus deltoides, Quercus macrocarpa, Q. rubra, Tilia americana,* and *Ulmus* penetrate essentially across the Province, and under pre-Columbian conditions these communities were rather well restricted to the fire-protected sides of rivers and to islands in the rivers (Fig. 119). Deciduous shrubbery, especially of *Corylus americana, Rhus glabra, Shepherdia argentea,* and *Symphoricarpos* formed a bordering fringe along the dry edge of the riparian strips, as well as occurring in patches in protected hollows of the topography elsewhere. Limestone bluffs along rivers supported a sprinkling of *Juniperus virginiana.*

A distinctive southern Section occurs as a belt along the Gulf Coast from central Texas to Tamaulipas.[227,252] *Andropogon scoparius* is here joined by *A. saccharoides, A. littoralis,* and other species of tropical affinities in *Cenchrus, Chloris, Eragrostis,* and *Paspalum,* usually with a scattering of *Prosopis juliflora* superimposed (Fig. 120).

Along the western margin of the *Andropogon scoparius* Province it forms an ecotone with the *Bouteloua gracilis* Province. Here it penetrates farthest into the drier steppe on stony loams, on subirrigated but nonsaline bottomlands, or on lower north-facing slopes that compensate for aridity in still

Fig. 119. *Quercus macrocarpa* in a ravine, with meadow–steppe of the *Andropogon* Province on uplands. The white-foliaged plant in the right foreground is *Artemisia frigida*. Near Marvin, South Dakota.

Fig. 120. *Prosopis juliflora* with *Opuntia* and *Andropogon* and other grasses, north of Laredo, Texas.

another way. To the east it abuts the *Quercus falcata* Province with large strips of *Quercus marilandica–Q. stellata* savanna penetrating deeply into the *Andropogon* Province as edaphic climaxes on sandy soils. From Oklahoma to Minnesota the Province is bordered by fire-maintained grassland that occupies land with forest potentialities. On the north the contact is with the *Festuca scabrella* Province.

Under primeval conditions major vertebrates of the midcontinental steppe Provinces included bison (in vast herds!), wapiti, pronghorn antelope, jackrabbit, prairie dog, ground squirrel, pocket gopher, grizzly bear, gray wolf, coyote, badger, kit fox, rattlesnake, sharp-tailed grouse, prairie chicken, and meadow lark.

Growing grain without irrigation is the major land use in the Province. Spring wheat, rye, oats, and barley are grown in the Dakotas. Maize, grown throughout, becomes most prevalent centrally, with winter wheat and sorghums added from Nebraska southward.

The *Festuca scabrella* Province[36,85,266,298]

The *Festuca scabrella* Province occurs along the southern border of the *Picea glauca* Province where it abuts steppe. It is a mosaic of grassland and *Populus tremuloides* groves (Fig. 121), with herbaceous vegetation on the convex surfaces and the groves related to depressions or protected slopes. Starting in southern Manitoba it forms a belt that curves northwesterly across Saskatchewan almost to central Alberta, then turns southward along the eastern flank of the Rockies to the northern edge of Montana (Fig. 55). A disjunct unit, nearly devoid of *Populus,* and with the low shrub *Potentilla fruticosa* regularly present, encircles the Cypress Hills near the southern boundary of Canada between Alberta and Saskatchewan.

Fig. 121. *Populus tremuloides* groves, east of Crookston, Minnesota.

The dark brown to black soils of the steppe matrix are outstandingly dominated by *Festuca scabrella* or *Stipa spartea curtiseta*, both rather large caespitose grasses which are accompanied by appreciable amounts of *Agropyron, Carex, Danthonia,* and *Koeleria cristata,* with *Festuca ovina* added westward. Since the Province borders forest and there is a substantial representation of forbs (*Achillea millefolium, Anemone patens, Antennaria, Cerastium arvense, Galium boreale, Geum triflorum, Solidago missouriensis,* etc.) the term meadow–steppe is appropriate for the herbaceous component of the zonal vegetation. Shrubs may be scattered singly, as are *Artemisia frigida* and *Potentilla fruticosa,* or they may be aggregated into floristically complex low thickets composed of *Elaeagnus argentea, Rosa,* and *Symphoricarpos occidentalis.*

Overgrazing favors *Danthonia parryi* (a large, palatable and productive perennial) near the Rockies, or *Artemisia frigida, Bouteloua gracilis,* and *Stipa comata* centrally, or *Poa compressa* and *P. pratensis* eastward.

Populus tremuloides, often with the taller *P. balsamifera* or *P. trichocarpa* appearing in the wetter centers of the groves, forms dome-shaped groves that are the largest and tallest, and occupy most of the landscape, as they approach forests of conifers at the northern border of the groveland belt. These groves tend to invade the encircling grassland by means of suckers from the shallow roots of *Populus tremuloides,* with this tendency held in check by desiccating winds, grazing, and fire. At times fire will run completely through a grove, girdling the trees, but root suckers appear promptly and soon overtop *Epilobium angustifolium* and other temporary invaders. Rainfall is relatively high for steppe (Fig. 12L), and the dark, fertile loams in the grove habitats remain moist all summer.

Wisconsin ice covered all of this Province, and many depressions left in the undulating till plain contain ponds or marshes supporting *Spartina pectinata, Carex,* and *Glyceria grandis* surrounded by *Salix* then by *Populus.* Whether the groves contain a central pond or marsh or not, they are streamlined with *Populus tremuloides* extremely dwarfed about the periphery (Fig. 12I). In these margins *Symphoricarpos* is usually conspicuous, with the taller shrubs in the interior including *Amelanchier alnifolia, Cornus stolonifera, Corylus, Elaeagnus argentea, Rosa, Rubus, Shepherdia canadensis,* and *Viburnum edule.* In the herbaceous strata are *Aralia nudicaulis, Cornus canadensis, Heracleum lanatum, Maianthemum canadense, Osmorhiza occidentalis, Pyrola asarifolia,* and *Smilacina stellata.*

Floodplain forests consist principally of *Acer negundo, Fraxinus pennsylvanica,* and *Ulmus americana.* The vegetation of tracts of sandy outwash is characterized by *Arctostaphylos uva-ursi, Calamovilfa longifolia, Juniperus horizontalis, Oryzopsis hymenoides, Prunus besseyi, Sporobolus cryptandrus,* and *Stipa.* Bogs are essentially identical with those of the *Picea glauca* Province.

Approaching the southerly margin of the Province the groves are small, widely scattered, and most strongly restricted to habitats of above-average moisture. Usually stands of *Festuca scabrella* occur well beyond the limits of the groves in this ecotone, but at the eastern end in northwestern Minnesota the groves extend farther into an ecotone with the *Andropogon scoparius* Province. The central and western portions of the groveland arc make contact with the *Bouteloua gracilis* Province. The major dominants of both the groves and the grassland matrix have floristic affinities with the north, which is in marked contrast with the affinities of the major dominants of the two steppe Provinces which abut the groveland belt on the south.

Most of the well-drained land in the *Festuca scabrella* Province has been converted to grain fields, with the remainder used for grazing.

The *Bouteloua gracilis* Province[419]

The *Bouteloua gracilis* Province, largest of the North American steppe Provinces, occupies the driest part of the rain shadow of the Rockies and extends from southern Alberta and Saskatchewan to Mexico (see Fig. 55). Since the floristic derivation of most of the dominant perennial grasses (*Aristida*, *Bouteloua*, *Buchloe*, *Hilaria*, *Muhlenbergia*, *Sporobolus*, and *Stipa*) links this Province with the Madro–Tertiary Geoflora, it stands clearly separated from both the *Andropogon scoparius* and the *Festuca scabrella* Provinces bordering it on the east and north, for these have affinities to the southeast and north, respectively. Furthermore, the vegetation of this Province lacks the character of meadow–steppe that is shared by the other two.

A northern Section of the *Bouteloua* Province, extending from Canada to Colorado, is distinguished by *Bouteloua gracilis* dominating a rather continuous layer of short graminoids, above which there is usually a more diffuse layer of taller grasses (Fig. 122). Dwarf shrubs are inconspicuous on the zonal soils; medium and tall shrubs are lacking.[70,81,170]

Rainfall here is as low as 250 mm a year in much of the area (see Fig. 13A). Normally this moistens the loams less than ½ m deep and closely restricts grass activity to the rainy season in early summer. Soil profiles belong to the Chestnut, Brown, or Sierozem Great Soil Groups, with the carbonate layer rising progressively higher toward the west where it may lie within 20 cm of the surface.

On these zonal soils the rhizomatous *Bouteloua gracilis*, usually mingled with xerophytic species of *Carex* such as *C. eleocharis* and *C. filifolia*, or with the stoloniferous *Buchloe dactyloides*, forms a conspicuous lower vascular stratum with foliage mostly less than 1 dm tall. Over most of the Section the upper vascular layer, with inflorescences rising up to about ½ m high, is variously composed of *Stipa comata*, *Agropyron smithii*, and *Koeleria cristata*. In the northern part of the Section *Agropyron dasystachyum*

Fig. 122. Steppe with an upper layer of *Stipa comata* and a lower layer of *Bouteloua gracilis*, near Three Forks, Montana.

and *Stipa spartea curtiseta* are additional components of the upper stratum, and in the southern part of the Section this layer may be represented by *Bouteloua curtipendula* or caespitose ecotypes of *Andropogon scoparius*. Here in the south *Buchloe* becomes more conspicuous relative to *Bouteloua* and *Carex*, and the taller layer is not as well developed as in the northern part of the Section. In years of abnormally low rainfall the taller grasses remain sterile and are proportionately more dwarfed, so that the vegetation appears essentially one layered.

Vasculares other than graminoids are relatively inconspicuous, these including forbs such as *Liatris, Malvastrum coccineum, Oxytropis, Petalostemon, Psoralea tenuiflora,* and *Solidago missouriensis,* and chamaephytes such as *Artemisia frigida, Eurotia lanata, Opuntia, Phlox hoodii,* and *Selaginella densa.* Northward the last may form a mat next the ground, especially on convex topography.

When the first Caucasian exploring parties came to the *Bouteloua* Province after crossing the relatively lush *Andropogon* Province during its spring flush of growth, the shorter and now-drying grasses they encountered caused them to report back that they had entered "The Great American Desert." This appellation soon fell into disuse as settlers arrived and began raising cattle and wheat, but when a series of wet years and high grain prices prompted widespread wheat growing that was then followed by below-normal rainfall in the 1930s, extensive wind erosion of the abandoned fields brought the term back into popular use for a time.

The native ungulates of the *Bouteloua* Province that were replaced by cattle during the 19th century were mainly bison, pronghorn, and wapiti, with deer wherever there was appreciable gallery forest. Although there were phenomenally large herds of bison roaming these plains when white man began intruding, the herds were constantly on the move so the vegetation had ample time to recover after each episode of close grazing. Heavy grazing, which started when white man introduced cattle and restricted their movements with fences, is far more detrimental to the taller grass layer, presumably since animals graze all species to about the same level and this takes a greater proportion of the shoots of the taller species. Both grass layers send roots to about the same depth, well over a meter at the wetter edge of the Province. Owing to the limited loss of foliage by the shorter grasses, and their greater root/shoot ratios, these species are remarkably tolerant of grazing, but they too can be overgrazed and replaced by the unpalatable *Aristida longiseta* or *A. purpurea*, or by shrubs such as *Artemisia frigida*, *Chrysothamnus*, or *Gutierrezia sarothrae*, or under extreme abuse by *Opuntia polyacantha* (Fig. 123).

Ever since its origin in mid-Cenozoic time the grassland east of the Rockies has developed under the influence of grazing ungulates. Horses and camels appeared early and continued to dominate until Asiatic grazers reached the area in Pleistocene time, then the older fauna dwindled to extinction leaving the Great Plains mainly to bison of Asiatic derivation by the start of the Hypsithermal Interval.[333] During the zenith of this dry period the *Bouteloua* Province apparently became intolerably dry for both the aborigines and their prey. Hunter and hunted moved up the slopes of the Rockies as aridity intensified, until they were living at altitudes now supporting alpine tundra.[206] By inference, the area now occupied by the

Fig. 123. Excessively grazed area near the *Stipa–Bouteloua* stand shown in Fig. 122, with the grasses replaced by *Opuntia polyacantha* and seedlings of *Salsola kali*. Under primeval conditions several large areas of *Opuntia* growing at least this dense occurred elsewhere in the northern Section of the *Bouteloua* Province, under circumstances where the dominance of this cactus could not be attributed to ungulate influence. Usually grazing must be extremely severe to bring about much of an increase of this plant.

Bouteloua Province may have been desert at that time. In view of the stress of that dry period it is remarkable that relic populations of trees that invaded the present *Bouteloua* Province during Wisconsin glaciation managed to persist there, as illustrated by *Quercus macrocarpa* and *Picea glauca* which survived in the Black Hills of South Dakota, and by *Picea engelmannii* which became stranded in the Bear Paw and Sweetgrass Hills of Montana.[107]

Prominent among the nonzonal types of vegetation in this Section in modern times are the thin and discontinuous gallery strips consisting mostly of shrubby *Salix* or arborescent *Populus angustifolia* or *P. sargentii* (Fig. 124). Swales that are somewhat protected from the wind and so accumulate above-average amounts of the meager snowfall may support a low deciduous scrub composed of such plants as *Elaeagnus argentea, Prunus, Ribes, Rosa,* and *Symphoricarpos occidentalis.* Flat areas, or slight depressions that accumulate clay and drain slowly after each convectional storm in summer, are characteristically dominated by the rhizomatous *Agropyron smithii*, the caespitose *Stipa viridula*, or the *stoloniferous Buchloe dactyloides* (Fig. 125). Where drainage is so poor that salts (mostly sodium sulfate) accumulate, the halophytic communities include *Atriplex canescens, A. nuttallii, Distichlis stricta, Hordeum jubatum, Puccinellia nuttalliana, Salicornia rubra, Sarcobatus vermiculatus, Scirpus paludosus, Spartina pectinata, Sporobolus airoides,* and *Suaeda depressa.*[397]

Protected slopes and stony soils with high infiltration capacity support outliers of the *Andropogon scoparius* suite of communities. In the southern part of the Section strips of *Andropogon gerardi* follow moist bottomland across even the driest area. In the northern part of the Section *Artemisia*

Fig. 124. Gallery forest in *Stipa comata–Bouteloua gracilis* steppe, with *Populus sargentii, Salix,* and *Shepherdia.* Little Big Horn River, Montana.

Fig. 125. *Buchloe dactyloides* spreading vegetatively over a shallow basin that holds water briefly after a heavy rain, near Amarillo, Texas.

cana commonly covers moist sandy stream terraces. *Juniperus scopulorum* is common at the summits of river bluffs.

In a large area (over 4 million ha) of stabilized Pleistocene dunes in western Nebraska, community types are dominated by *Andropogon hallii, A. scoparius, Bouteloua gracilis, Calamovilfa longifolia,* and *Stipa comata* (Fig. 126).[59] Other sandy tracts of slightly different character support edaphic climaxes in which *Artemisia filifolia, Calamagrostis gigantea, Oryzopsis*

Fig. 126. *Calamovilfa longifolia* and *Sporobolus cryptandrus* dominant in the large area of sandhills in Nebraska.

hymenoides, Redfieldia flexuosa, Sporobolus cryptandrus are the major species. Locally, *Ipomoea leptophylla* or the stemless *Yucca glauca* become conspicuous on these sandy soils.

In addition to the bison, wapiti, pronghorn, and deer mentioned previously, other important herbivores included big horn sheep, beaver, hare (jackrabbit), and the colonial prairie dog. The grizzly bear, gray wolf, badger, ferret, and coyote were the major carnivores.

In central Kansas and the cool northern parts of the Section dryland farming is moderately productive of wheat, but wind erosion is always a problem, and years with below-average precipitation are economically extenuating. Natural succession on abandoned fields leads to an approximation of the zonal climax in as short a time as 40–100 years.[427]

A southeastern Section of the *Bouteloua gracilis* Province, extending from southern Colorado to northern Arizona then south to Chihuahua, is likewise dominated by *Bouteloua gracilis*, but here it is accompanied by a variety of associates with differing statures (*Aristida longiseta, Bouteloua curtipendula, B. hirsuta, Hilaria jamesii, Muhlenbergia,* and *Stipa*) which do not provide the appearance of a distinct overstory as in the north.[83,255,305] Also there is usually a conspicuous scattering of a single species of shrub in each association, such as *Acacia, Lycium pallidum, Nolina texana, Opuntia, Koeberlinia, Prosopis juliflora,* or *Yucca.* Cold winters still enforce dormancy of this vegetation.

The compositional changes induced here by excessive grazing pressure favor other grasses (*Schedonnardus paniculatus, Sitanion hystrix*) and shrubs (*Chrysothamnus, Gutierrezia sarothrae, Opuntia*) or the annual weed *Salsola kali.*

Sandy soils support *Andropogon hallii, Artemisia filifolia,* and *Sorghastrum nutans.* Communities of low trees on stony soils include *Juniperus* or *Quercus havardii.* Nonsaline swales are typically dominated by *Hilaria mutica* or *Buchloe dactyloides,* whereas salinity favors *Atriplex canescens.* Streamsides support little more than *Salix.*

From central New Mexico to central Arizona and southward into Mexico, there is a distinctive southwestern Section of the *Bouteloua gracilis* Province that occurs at lower and warmer elevations, but with slightly more rainfall. Vegetative activity is keyed primarily to the summer rainy season (Fig. 13B).

Bouteloua eriopoda is the characteristic dominant, with common associates including additional species in that genus, especially *B. rockrothii, Andropogon, Aristida, Eragrostis intermedia, Hilaria belangeri, Muhlenbergia, Sporobolus, Trichachne,* and *Tridens pulchellus.*[305,425] Scattered shrubs 1–6 m tall occurring singly also characterize this southwestern Section: *Calliandra eriophylla, Dasylirion wheeleri, Ephedra trifurca, Flourensia cernua, Koeberlinia spinosa, Larrea tridentata, Nolina, Opuntia, Prosopis juliflora,* and *Yucca* (Fig. 127).

Fig. 127.   Steppe dominated by *Bouteloua* and *Sporobolus airoides*, and dotted with *Nolina microcarpa*. Near Sonoita, Arizona.

The *Prosopis* is also characteristic as a dominant of sandy soils,[62] unless the sand is composed of gypsum.[127] Trees that occur in deep ravines with permanent streams are *Acer negundo, Juglans rupestris*, and *Platanus wrightii*, but shallow drainageways with intermittent water are lined with no more than shrubs: *Acacia greggii, Baccharis, Brickellia laciniata, Chilopsis linearis, Fallugia paradoxa, Hymenoclea monogyra, Rhus microphylla*, and *Sapindus drummondii*, etc. Shallow basins that are slow to drain may be covered *Hilaria mutica* or *Sporobolus wrightii*.

Paired photographs show conclusively that in many places in both the southeastern and southwestern Sections the woody plants that once dotted the steppe, especially *Prosopis juliflora*, have thickened to form a continuous stratum, while the grasses have dwindled to insignificance. In one interpretation, heavy grazing when cattle replaced the sparse population of native prong horn and mule deer so thinned the natural grass cover that it could no longer carry fire, then the fire-sensitive shrubs which had been kept in check by occasional natural fires increased in density.[204] In this interpretation at least part of the Province formerly had the status of a fire climax.

An alternative interpretation of the replacement is that since in the past century the climate has become drier, the shrub increase represents an expansion of desert.[181] Weather data and the mortality of desert plants in areas free from grazing and fire tend to support this hypothesis.

Practically the only use of the land in both the southeastern and southwestern Sections is cattle grazing. For this the vegetation is very well suited, since the grasses are nutritious even when cured, and can be grazed the year around. *Prosopis* pods are sweet and nutritious, and were an important item of food for the aborigines. Livestock too eat the pods avidly, and honey from the flowers is of outstanding quality.

In the mosaic type ecotones, *Bouteloua* grassland alternates with communities of chaparral or encinal above and desert below, with grassland occupying the relatively heavy soils and woody vegetation the coarser materials.

The *Agropyron spicatum* Province

In Miocene time the area from central British Columbia to the northern edge of the Great Basin was occupied by temperate mesophytic forest, with chaparral continuing southward. When the Cascade–Sierra axis uplifted in the succeeding epoch, rainfall declined to the lee of these mountains until nearly all the above woody vegetation was replaced by steppe. The chief plants to preempt this newly arid trough (*Agropyron spicatum, Festuca idahoensis, Poa, Artemisia,* and *Eurotia*) were of boreal extraction, and even during Pleistocene glaciation this steppe vegetation was not completely eliminated. During the Hypsithermal Interval that followed, forest was pushed back upslope above even modern limits and a moderate infusion of austral species of Madro–Tertiary affinities took place, these including *Eriogonum, Grayia spinosa, Sarcobatus vermiculatus,* and *Tetradymia canescens.* Even during this period of maximum aridity the climate never became dry enough to eliminate a grass cover and sponsor desert.

In this Province precipitation falls mainly in winter (Fig. 13C,D), and in the northern half especially, the winters are mild enough that aestivating grasses put forth new leaves in autumn and grow fitfully all winter. Everywhere there is a period of maximum vegetative and flowering activity in spring, followed by summer desiccation of the environment and aestivation of the herbs. Only deep-rooted shrubs retain green leaves throughout summer.

Precipitation ranging from 150–500 mm a year permits a distinct zonation, especially northward where the most xerophytic vegetation in the lowest and driest areas may be surrounded by a series of increasingly less arid Zones terminating in meadow–steppe next the forested foothills. Along this moisture gradient soils range from Sierozem to Brown to Chestnut to Chernozem to Prairie Podzolic.

Two Sections of the *Agropyron spicatum* Province (Fig. 55) may be recognized. In the northern Section, from central Oregon northward, there is a strong oceanic element in the climates. Here a strip of shrub–steppe extends in a north–south direction along the lowest, hottest, and driest part of the trough between the Cascades and Rockies.[104,272,401] The tallest plant, *Artemisia tridentata tridentata,* usually accompanied by small amounts of

Fig. 128.   Virgin shrub–steppe (*Artemisia tridentata tridentata–Agropyron spicatum* association) in south central Washington. Canopy coverage of *Artemisia* here was 9%, *Agropyron*
44%.

Chrysothamnus viscidiflorus, Grayia spinosa, or *Tetradymia spinescens,*
forms a monotonous layer of wide-spaced gray shrubs 0.5–2.0 m tall that
covers 5–20% of the ground (Fig. 128). Beneath and between these shrubs is
a tall (3–4 dm) herbaceous layer consisting chiefly of caespitose grasses*
(*Agropyron spicatum, Festuca idahoensis, Stipa thurberiana*), and still lower
is a layer of the tiny perennial *Poa sandbergii* with an abundance of equally
short winter annuals. The few forbs include *Balsamorrhiza, Calochortus,
Astragalus, Castilleja, Erigeron,* and *Lomatium.* Since all grasses and forbs
aestivate, and the shrubs burn readily even though in full leaf in summer,
fires are common (Fig. 129). The perennial herbs and most shrubs regenerate directly from subterranean organs, but fire kills the *Artemisia tridentata*
and it must reenter a burn by dissemination of its achenes.

Adjacent to the shrub-steppe there is usually a physiognomically distinct
belt of steppe differing principally in the absence of *Artemisia* and the substitution of *Chrysothamnus nauseosus* for *C. viscidiflorus.*

*Shantz[359] took many photographs in the *Artemisia* belt, but all appear to have been made at
railroad stations where livestock had eliminated the grass while awaiting their shipment, consequently, he made no mention of the heavy grass layer of natural vegetation in his report.
Shreve,[365] who included this Province with "desert" also had an astonishingly inadequate
concept of the grass component of the vegetation in its ungrazed condition, and of the geographic extent of this rather homogeneous unit. Weaver and Clements,[420] on the other hand,
hypothesized that grassland devoid of shrubs was the potential vegetation of this entire Province, except for a limited area extending from central Utah to southwestern Idaho, northeastern
California, and northern Nevada!

Fig. 129.   In the foreground fire has eliminated the shrub *Artemisia tridentata tridentata*, leaving the perennial grasses, chiefly *Agropyron spicatum*, stimulated. Near Sunnyside, Washington.

Finally, just before moisture becomes adequate for forest, there is a belt of meadow–steppe in which the *Agropyron spicatum* (in rhizomatous form here) and *Festuca idahoensis* are joined by a profusion of perennial forbs (*Balsamorrhiza sagittata, Calochortus, Castilleja, Geranium viscosissimum, Helianthella uniflora, Lupinus, Potentilla*) and a few shrubs (*Artemisia tripartita, Eriogonum heraceloides, Purshia tridentata, Rosa, Symphoricarpos albus*) (Figs. 130 and 131). The complementary distributions of the shrubs is the most useful criterion for distinguishing associations in different parts of this meadow–steppe fringe.

In the southern Section, extending from central Oregon and southern Idaho southward, the influence of the Westerlies is not so strong, and the climatic climaxes are shrub–steppe practically throughout with structure and composition closely similar to the *Artemisia tridentata* shrub–steppe component of the northern Section (Fig. 132). Here in places *Agropyron dasystachyum* becomes an additional important grass, and in places *Peraphyllum ramosissimum* or *Ephedra* make minor contributions to the shrub stratum.

Saline soils, which are especially widespread in the southern Section owing to the evaporation of Pleistocene lakes Bonneville (whose modern remnant is Great Salt Lake) and Lahontan, support a rich flora of halophytes including *Chenopodiaceae* (*Allenrolfea occidentalis, Atriplex, Dondia, Kochia, Salicornia,* and *Sarcobatus vermiculatus*) and Gramineae (*Distichlis stricta, Elymus cinereus,* and *Sporobolus airoides*) (Fig. 133).[140] Nonsaline seasonal marshes once supported widespread communities distinguished by an

Fig. 130. Virgin meadow-steppe in southeastern Washington (*Festuca idahoensis–Symphoricarpos albus* association), with low deciduous thickets of *Symphoricarpos albus, Rosa,* and *Prunus virginiana melanocarpa.*

Fig. 131. Detail of the herbaceous matrix in the steppe mosaic shown in Fig. 130. Conspicuous forbs in the photo are *Balsamorhiza sagittata, Helianthella uniflora, Castilleja cusickii, Lithospermum ruderale, Haplopappus liatriformis, Astragalus palousensis,* with *Symphoricarpos albus* in the lower left corner. The major grass is *Festuca idahoensis.*

Fig. 132. *Populus angustifolia* fringe closely confined to a stream margin in shrub–steppe dominated by *Artemisia tridentata* and *Stipa comata*. Near Dubois, Wyoming.

Fig. 133. Saline depression with a *Distichlis stricta* belt rimmed by a narrow fringe of *Sarcobatus vermiculatus*, then with *Artemisia tridentata* shrub–steppe on nonsaline uplands beyond. Grand Coulee, Washington.

abundance of *Camassia quamash,* the farinaceous bulbs of which provided the major plant food of the aborigines.

Edaphic climaxes on sands are characterized by *Stipa comata* associated with scattered shrubs of *Artemisia tridentata, Chrysothamnus,* and *Tetradymia.* In the southern Section other special soils such as lithosols and those with shallow caliche, are dominated by dwarf shrubs such as *Artemisia arbuscula, A. nova, A. rigida, A. spinescens,* and *Eurotia lanata,* each associated with a grass layer (Figs. 134 and 135). The northern Section contains a rich suite of communities on lithosols that have dwarf shrubs in *Eriogonum* or *Artemisia rigida* associated with a *Poa sandbergii* layer.

In the meadow–steppe of eastern Washington and adjacent Idaho and Oregon, *Crataegus douglasii* formed tall dense thickets on stream terraces and northerly slopes. Gallery forests consist mainly of narrow strips of *Populus trichocarpa* (northern Section) or *P. angustifolia* (southern Section). Groves of *P. tremuloides* are common on permanently moist spots in the least arid parts of this Province, sometimes covering considerable area, and on the eastern margin of the southern Section they form an almost continuous belt along lower timberline.

Where mountain ranges rising within the intermountain trough are either not massive enough or are improperly oriented to stimulate the usual amount of increase in precipitation with elevation, vegetation dominated by *Artemisia tridentata vaseyana* with *Festuca idahoensis, Stipa columbiana,* and other grasses continue steppe vegetation far upslope, in places extending so far as to form an ecotone with alpine tundra.

Fig. 134. Ecotone between the *Artemisia tridentata–Agropyron spicatum* association on deep soil at left, and the *Artemisia rigida–Poa secunda* association at right on a lithosol. Central Oregon.

Fig. 135. Ecotone between an *Eurotia lanata* community in the foreground, and a strip of *Artemisia tridentata* along a now dry drainage channel. Wells, Nevada.

In immediately pre-Columbian times there were very few large grazing animals in this Province—only a scattering of pronghorn with temporary small incursions of bison entering by way of the discontinuity of the Rockies in Wyoming. Occasional heavy winter snows eliminated each wave of bison immigration, for these animals are adapted to areas where limited winter precipitation leaves steppe grass available at least on convex topography. There were abundant hares (jackrabbits), cottontail rabbits, ground squirrels, sage hens, and sharp-tailed grouse which were preyed upon by coyote, bobcat, and badger.

Grazing is now a major land use in areas where irrigation water is not available, or where the zonal soils are too interrupted for agriculture, but the native grasses tolerate light use only. This has been thought to be a consequence of weak grazing pressure during the milennia during which the species were developing adaptations to their total environment. Modern grazing disclimaxes are conspicuous and mostly dominated by aliens. In meadow–steppe and *Populus tremuloides* groves where *Rosa* and *Symphoricarpos* abound, *Poa pratensis* increases to the point of dominance under continual heavy grazing. Elsewhere the increasers are mainly annuals—*Bromus tectorum*, *Elymus caput-medusae*, or in just the southern Section, the stock-poisoning *Halogeton glomeratus*. *Chrysothamnus* and *Artemisia tridentata* usually increase in these zootic climaxes. Wherever overgrazed land can be cultivated, the alien *Agropyron cristatum* or *A. desertorum* have been planted, as these Asian steppe grasses withstand grazing

pressure quite well. Feral horses are multiplying rapidly and aggravating the overgrazing problem.

Annual crops of wheat and peas yield well in unirrigated meadow–steppe in the northern Section, but in drier zones hardly more than wheat can be grown, and an alternate year of fallowing is necessary. In the hottest and driest areas where *Artemisia tridentata* shrub–steppe occurs, irrigation permits highly diversified agriculture, producing fruits, vegetables, grains, sugar beets, and white potatoes, etc.

The *Agropyron spicatum* Province is in contact with the *Quercus gambelii*, *Pinus cembroides*, *Pinus ponderosa*, or *Pseudotsuga* Provinces on the slopes of the surrounding foothills, depending on the degree of upward truncation of these ligneous Provinces. In central Wyoming a discontinuity in the Rocky Mountain chain allows this Province to extend through the mountains and make contact with the *Bouteloua gracilis* Province at the western margin of the Great Plains. This steppe ecotone corresponds with the change from predominately winter to predominately early summer precipitation. Across the floor of the Great Basin in central Utah and Nevada the *Agropyron spicatum* Province forms an ecotone with the *Atriplex confertifolia* Province to the south (see Fig. 55), but by ascending a short distance up on the foothills just above the floor of the Basin, slender extensions of the former continue as far as the southern extremity of the *Atriplex* Province.

The *Atriplex confertifolia* Province[33,135,423]

Zonal soils of the Great Basin immediately south of the *Agropyron spicatum* Province support a shrub–steppe that appears as a monotonous expanse of low shrubs, well spaced, the principal species of which is *Atriplex confertifolia* (Fig. 136). Accompanying shrubs, mostly of similar stature, include *Artemisia spinescens*, *Ephedra nevadensis*, *Grayia spinosa*, *Lycium cooperi*, and *Sarcobatus baileyi*. These shrubs are mostly spiny and gray-leaved, and are approximately 3 dm tall. The perennial grasses, which are also smaller than corresponding elements of the *Agropyron* Province, include *Oryzopsis hymenoides* and *Hilaria jamesii* with lesser amounts of *Bouteloua gracilis*, *Sitanion hystrix*, and *Stipa*. Winter annuals, especially those in the genus *Eriogonum*, are quite abundant in this climate which has relatively dry summers (Fig. 13E). Where the soil is disturbed the alien annuals *Salsola kali* or *Halogeton glomeratus* are common. Although grass is conspicuous in areas not depleted by overgrazing, the vegetation is too open to carry fire except in the hottest and driest weather. The openness of the grass layer and the dwarfness of both shrubs and grass reflect an approach to desert.

The soils, technically Gray Desert Soils, may be salt free, or saline to within 30 cm of the surface. Some are also alkaline throughout. Owing to thin vegetation and consequent deflation, the surface is generally sandy with a "desert pavement" of wind-polished stones.

Fig. 136. *Atriplex confertifolia* with *Hilaria jamesii, Aristida longiseta;* and *Opuntia.* Near Wellington, Utah.

Characteristic dominants of dunes include *Artemisia filifolia, Chrysothamnus stenophyllus, Dalea polyandenia*, and *Tetradymia*. Saline depressions are populated by *Allenrolfea occidentalis, Atriplex nuttallii, A. corrugata, Distichlis stricta, Kochia vestita, Sarcobatus vermiculatus, Sporobolus airoides*, and *Suaeda fruticosa*. Other types of soil or topography support communities dominated by *Artemisia nova, Coleogyne ramosissima* (Fig. 137) or *Eurotia lanata*.

Cultivation is not practical in this Province and the land is used mainly for cattle grazing. Animal life resembles that of the *Agropyron* Province, with the kit fox added.

The *Atriplex* Province is surrounded on the north, east, and west by the *Agropyron spicatum* type of steppe with *Artemisia tridentata*. To the south it forms an ecotone with the *Larrea* Province of the Desert Region (Fig. 55).

The *Stipa pulchra* Province

Prior to Caucasian influence the floor of the Sacramento–San Joaquin Valley in California, as well as the western foothills of the Coast Ranges southward from San Luis Obispo, supported steppe. But cattle and horses introduced centuries ago by Spaniards were allowed to multiply in a semiwild state to tremendous numbers, and this led to an essentially complete destruction of the native perennial dominants and their replacement by alien annuals. Occasional stands of grassland dominated by native perennials (*Stipa pulchra*, variously associated with *S. cernua, S. coronata, S. lepida, Elymus triticoides, Aristida divaricata, Poa scabrella, Koeleria cris-*

Fig. 137. Stand of *Coleogyne ramosissima* on very sandy soil. House Rock Valley, Arizona.

tata, Melica californica, and *Muhlenbergia ringens*) have usually been interpreted as approximations of the pre-Caucasion climaxes. Perennial forbs appear to have been well represented, along with native annuals in *Aristida, Eragrostis, Escholtzia, Lotus,* and *Orthocarpus.* The observed tendency for *Stipa pulchra* to increase when given protection from cattle and horses tends to strengthen the concept of its major role in this steppe Province, but at the same time this grass is favored by burning, so its former role is not clear.

The principal alien annuals which now dominate are *Avena barbata, A. fatua, Bromus mollis, B. rigidus, B. rubens, Erodium, Festuca megaleura, Hystrix patula, Medicago hispida,* and *Trifolium.*

Owing to the cool-winter and hot-dry-summer climate (Fig. 13F), vegetative activity starts with the first rains in autumn, slows during the frosty winter, then allows rapid maturing in spring before drying up about the first of May.

Chaparral scrub, which is climatic climax in adjacent areas, occurs as edaphic climaxes on rock outcrops in this steppe area (Fig. 138), and strips of forest, savanna, or groveland, mainly dominated by *Quercus lobata,* bordered the larger rivers up to 10 km in width. Strictly riparian trees here include *Fraxinus latifolia, Juglans hindsii, Platanus racemosa, Populus fremontii,* and *Salix.*

The low central part of this great valley was so nearly flat that it contained well over a million hectares of seasonal marsh and fen before these were drained for agricultural use. *Carex, Eleocharis, Scirpus,* and *Typha* were the major dominants here. Toward the southern end of the valley rainfall was inadequate to flush away salts, and the saline soils there supported expanses

Fig. 138.   Chaparral as an edaphic climax in annual grassland that has replaced *Stipa pulchra* steppe. Mt. Diablo, California.

of *Allenrolfea occidentalis, Distichlis stricta, Dondia moquini, Sarcobatus vermiculatus, Sporobolus airoides,* and *Atriplex.*

In the primeval state this Province supported herds of wapiti, pronghorn, mule deer, and hare with grizzly bear, cougar, and wolf as major predators. The condor, a scavenger now almost crowded to extinction in the south, once soared over all this steppe.

Parts of the valley that can be irrigated are now used intensively for diversified agriculture, including fruit and nut culture, with the other land used for grazing. The northern part of the Province is especially suited for the growth of rice.

THE DESERT REGION[306]

Desert has been defined in many ways. Geographers commonly define it strictly in terms of climatic data. To some it is any barren area irrespective of climate. Here it will be defined as an area with low precipitation in relation to the heat level, so that the solum is moist for only brief periods. Also, whatever vegetation occurs on zonal soils is dominated by shrubs, with herbs represented mainly by annuals. Negligible representation of perennial grasses on zonal soils is a key character.

Despite the physiognomic dominance of a single life form, i.e., shrubs, desert landscapes vary considerably from place to place. The least arid portions of the Desert Region support tall, nearly closed communities, whereas in drier areas the shrubs are short and widely spaced. In still more stressful climates the shrubs become confined to channels that carry flash floods, and in the most extreme deserts there is almost no vascular vegetation.

In some communities a single growth form, if not a single species, makes up nearly all the vegetation. In other communities there is a surprisingly wide variety of growth forms. Apparently in the absence of sufficient water to support dense vegetation that would result in severe competition for light, evolution has been more free to take different pathways in solving the problems posed by physical factors.

About 50% of the vascular desert flora is composed of therophytes, with most of the other species divided about equally among geophytes, hemicryptophytes, chamaephytes, and nanophanerophytes. In the deserts of southern California only about 25% of the flora is polyploid—a percentage well below that of other major vegetation types.[219] Since therophytes are considered relatively advanced types, it is curious that the primitive diploid condition should prevail.

Where there is a summer rainy season a special group of annuals germinate after heavy rains then complete their life cycles in 6–8 weeks before the soil is dry again, thus spending most of the year as dormant seeds. These are aptly called summer annuals or ephemerals, but in one sense they are not necessarily annuals, since the seeds may remain dormant for several years if rainfall remains below normal. They commonly keep their stomata open throughout the day—a habit which enables them to use the supply of solar energy to the fullest during the brief period when moisture is available to them. Since the shortening of their life cycles to fit the abbreviated rainy seasons is their chief adaptation, these annuals can be said to be "drought evading" plants. They are shallow rooted, and have no xeromorphic characters.

Where there is a rainy season in autumn or winter, the special group of annuals that germinate as these rains start have a longer period of development and grow larger, then flower in spring as the showers end and temperatures rise. These are the winter annuals, and they can hardly be considered ephemerals.

A notable attribute of the annuals (and other desert herbs as well) is the tremenduous variation in size of the individuals from year to year depending on the adequacy of the rains. Still another adaptation common among desert annuals is seed heterogeneity that results in some seeds responding promptly to rainfall with others requiring more time or more leaching away of inhibitors, thereby ensuring against the vagaries of weather devastating more than a part of the population.

A third life form that is conspicuous in only certain desert areas, but one that is so bizarre as to always attract much attention, is the succulent perennial. These plants have leaves or stems with abnormally thickened parenchymatous tissues which become turgid during rainy periods, then as the soil dries out the plants transpire very slowly, drawing on the water reserves in their tissues until there is another opportunity for replenishment. Such plants have a type of metabolism in which stomata remain closed during the day, with photosynthesis based on CO_2 taken from malic or other organic acids that accumulate while the stomata are open at night. Their shallow root systems enable them to benefit from even light rains, and some produce ephemeral rootlets during brief rainy seasons. In the Americas the stem-succulent *Cactaceae* together with leaf-succulents in *Agave* and *Dudleya* provide the bulk of the desert succulents. In Africa closely similar life forms are provided by the *Euphorbiaceae, Aloe, Gasteria, Haworthia, Lithops,* and *Mesembryanthemum.*

Since most annuals pass their life cycles quickly then disappear, and succulents are not found in most of the Desert Region, the nonsucculent shrub is the common denominator of desert physiognomy everywhere. These plants typically have extensive root systems that are either deep (to 53 m!)[320] and depend on rarely recharged subsurface moisture, or shallow and depend on light rains that are more frequent. The root:shoot ratio is commonly in the order of 6:1. For the most part, the leaf blades are very small (Fig. 139), which is important in preventing excessive heating under strong insolation, and the leaves are usually wholly or partly deciduous (e.g., *Encelia, Fouquieria*) and commonly xeromorphic and waxy (*Larrea*) or hairy (*Franseria*). Usually their stomata close during midday, and as the water supply becomes critically low they can reduce transpiration to a very low rate. Often the stems remain green for many years, so that complete or partial loss of foliage during the dry season does not prevent photosynthesis completely.

Fig. 139.   Foliage sprays and fruits of *Larrea divaricata* (left) and *Cercidium microphyllum.* Leaflets of *Larrea* are slightly less than 1 cm long.

The stems are commonly thorny. In most deserts around the world legumes contribute heavily to this shrub category.

Both succulent and the nonsucculent shrubs are "drouth minimizing" in that by various strategies their vegetative organs remain somewhat active during the dry seasons by making minimal demands on limited water supplies stored in either their tissues or deep in the subsoil. Still another category of desert plants with respect to their water relations consists of those that are "drouth enduring," in that their protoplasts dehydrate to extremely low levels without lethal consequences. In this category are lichens, bryophytes, and the remarkable *Selaginella lepidophylla*.

Those deep-rooted woody plants which tap water supplies far below the solum, especially along drainageways, may transpire even more vigorously than the average mesophyte. Some are so free of drouth stress that they flower regularly as the end of the dry season approaches.[395]

Succession is not conspicuous in deserts, since the communities are too open for shade elimination to operate. In fact, shade is a distinct boon to the seedlings of many species, and since it is essential for the survival of *Cereus giganteus*, that species cannot invade a denuded area until others enter and provide the necessary shade. Following denudation, commonly one or a few aggressive species of the climax invade and achieve dominance for a time before late-comers reduce their density, indicating that there are indeed competitive interactions. Otherwise there are probably slow changes taking place where progressive erosion or deposition are changing substratal conditions, or where lime, gypsum, or other salts are accumulating, or where there have been appreciable changes in climate over a few decades.[181]

Origin and Distribution

The general lack of suitable conditions for fossilization in arid regions, and consequent lack of a good fossil record, imposes severe limitations on any attempt to trace the history of vegetation in those regions. Inference must be drawn from such indirect evidence as the presence of ancient deposits of salts (borates, carbonates, sulfates) such as commonly result from the evaporation of water from basins, from the occurrence of ventifacts or a dark "desert varnish" on rocks, from alluvial fans of coarse conglomerate, from the relatively xerophytic character of fossils in semiarid land that is presumed to have been adjacent to desert, and even the near lack of fossils in itself cannot be ignored. Caves with stratified middens left by rat and sloth dung have shed light on desert history in southwestern United States, but the record extends back only about 12,000 years.

The two dry climatic belts centered on approximately 30° north and south latitude have probably favored some type of xerophytic vegetation throughout the history of land plants, but in North America there is no substantial

evidence that the Desert Region existed prior to the Pliocene Epoch.[18]
However, even in Eocene time, sediments in northeastern Nevada entrapped fossil remains of encinal, and drier environments probably existed to
the west and south where drouth would be expected to have been more severe.

Presumably exposed ridges with thin soil could have been fostering the
evolution of xerophytes in local topoedaphic climaxes since the dawn of
angiosperm history, so that preadapted plants were available to populate
climatically dry territory as it developed (Fig. 140). The highly specialized
xerophytes of our well-diversified modern desert floras do indeed suggest a
long period of evolution. The *Fouquieriaceae*, for example, is both highly
specialized and lacks close relatives, which certainly indicates an age greater
than Pliocene deserts.

Floristic affinities show that the North American deserts have drawn most
of their species from the Madro–Tertiary Geoflora. After the coalescence of
desert vegetation and its expansion in Pliocene time, glacial episodes of the
Pleistocene Epoch undoubtedly brought on repeated shrinkages with expansions intervening.[281] Just as alpine floras must have suffered when warm
interglacials extinguished their populations on all but a few of the highest
peaks, so pluvial climates south of the major ice caps probably purged desert
floras as they became fragmented then eliminated them in all but a few of the
lowest and driest areas that remained. However, in Coahuila there is no
evidence of significant change in xerophytic vegetation during that
Epoch.[287]

Fig. 140. *Opuntia* prospering on thin soil in seasonal forest on Big Pine Key, Florida.
Agave, Yucca and other xerophytes also occur nearby in special habitats.

Desert now occurs on the west side of each continent, centered roughly on about 30° latitude. This involves five major segments of the Desert Region, each of which has a high degree of floristic individuality: (1) southwestern United States and northern Mexico, (2) west of the Andes in Peru and northern Chile, and just east of the Andes in northern Argentina, (3) southwestern coast of Africa (Namib Desert), (4) a broad belt stretching from northern Africa across Arabia and eastward into Asia, and (5) northwestern Australia.

In addition to these major desert areas in the subtropical dry belt, rain shadows of mountains often extend desert well beyond these latitudes, as illustrated by the poleward extensions of desert into southern Nevada and to the north of the Himalayas, and the equatorward extension into Somaliland. Aridity along coasts that results from cold ocean water and stable air also extends desert longitudinally, as illustrated by the Humboldt Current flowing northward and extending desertic conditions nearly to the equator in Ecuador.

Many parts of the earth are commonly referred to as "desert," but do not fit the phytogeographic definition followed here. The *Bouteloua gracilis* Province was lumped with all the other unforested parts of western North America as "desert" in the early nineteenth century, and the intermountain parts of the Steppe Region still are often referred to as "desert." Such usage is not restricted to North America. The Kalahari "desert" of Africa, and central Australia, are both well supplied with perennial grass.

The subtropical desert belt is bounded on the poleward side by chaparral or temperate xerophytic forest, on the eastern side by steppe, and on the equatorward side by microphyllous woodland.

Environment

Arid climates (Table 3) are mainly products of the Subtropical High Pressure Belts (see Fig. 7) that lie between the Westerlies on the poleward side and the Trade Winds (Easterlies) on the equatorial side.[174] In these belts air masses are descending and dry, so that rainfall is minimal, relative humidity low, and insolation intense owing to the sparsity of clouds. These dry belts extend across oceans as well as continents.

The fullest expression of this aridity is at the western margin of each continent, with the stress diminishing gradually eastward in the same latitude, if mountains do not interfere. Typically this results in a sequence starting with desert at the western side of the land mass, followed by steppe or savanna, and ending with seasonal forest at the eastern margin. Since this dry belt is an astronomic phenomenon it has probably existed in both hemispheres throughout the earth's history, although the degree of its expression, the depth of its inland penetration, and its exact north–south position have varied from time to time. For example, changes in the configuration of the ocean bottom could affect ocean currents, with this in turn influencing the pattern of climates at least near the coast. Continental drift has also been of extreme importance in more remote Epochs and Eras.

Rainfall in this latitudinal belt is dependent upon incursions of moist air from adjacent Regions, descending as heavy convectional showers for the most part. The mean annual value varies from near 0 (on the west coast of South America between 20–26° latitude) to about 280 mm (as at Tucson, Arizona). Equally important as the meager supply, variability increases as the annual average decreases. The coefficient of variation may equal the mean! Thus, the average is more meaningful to the mathematician than to the ecologist. At Trujillo, Peru, only 35 mm of rain fell during 7 years, then in one month 388 mm fell! Such brief downpours are of limited areal extent, and commonly exceed the capacity of the mainly bare soils to take in the water so that much is lost as runoff, yet a single deluge raises the mean well above the median for many years.

Streams in deserts are intermittent unless their sources lie in wetter climates. Water courses are mostly short and stand empty most of the time, but during storms, runoff starts as sheets then becomes concentrated in drainageways that become raging torrents and may discharge water onto flat areas or into shallow basins which become temporarily ponded. Since water infiltrates deepest in the gravels of runoff channels, deep-rooted shrubs or small trees commonly form narrow gallery strips along the sides of these channels. Obviously, water is not equally scarce over a desert. Elsewhere soil moisture is deficient most of the year, and often for periods well exceeding a year. Rainfall usually penetrates but a short distance into the soil, with the subsoil moisture recharged only at long intervals. The few deep-rooted plants that achieve contact with the capillary fringe above a deep water table appear to represent individuals fortunate enough to have germinated in those exceptionally wet years when heavy rainfall wets the profile down to the capillary fringe.

Whether the storms come regularly at one or two seasons of the year, or at widely and unevenly spaced intervals, they trigger a pulse of activity in the vegetation that is proportionate to the water yield. Vegetation then lapses into a quiescent state pending the next rains. Light showers may stimulate activity in only the surface crust of lichens and algae. Successively heavier showers may benefit progressively more deeply rooted plants and stimulate seed germination. Daylength, the annual march of temperature, or the first shower of the rainy season are not the important triggering mechanisms in the desert that they are in climates with dependable wet seasons.

Nocturnal fog may be frequent in coastal strips of desert where there is an upwelling of cold water that renders the air too stable for convectional overturns, yet increases the relative humidity so that moisture condenses as fog and is drawn onshore at night. In such places as the Atacama Desert of Peru and the Namib Desert of southwestern Africa, the condensation of nocturnal fog along the coast provides most of the biotically useful supply of water. For the most part these fog deserts are coastal extensions of the subtropical desert belt.

The highest maximum temperatures recorded by weather observers are in the latitudinal desert belt. An air temperature of 56°C was recorded in a standard weather shelter in Death Valley, California. Heat is therefore another special stress that must be accommodated by desert organisms. Small leaves and light colored or hairy surfaces are common among desert plants, these serving to reduce their heat load. Clear skies also promote strong radiant loss of heat at night, so cold nights are teamed with hot days. However, in the coastal fog deserts the mean maximal temperatures can be surprisingly low. At Antofagasta, Chile, the value is 24.4°C.

Topography is very important in determining soil patterns over most of the North American Desert Region. The area is studded with mountains from which brief torrential rains wash gravel and boulders onto the margins of the broad intervening valleys. Close to the base of a mountain the coarsest materials come to rest, forming alluvial cones or fans with a decided slope to their surfaces. Finer materials are carried farther out and deposited as loamy gravels having a more gentle slope. Silt and clay are carried to the lowest elevations where they are either removed by streams, or accumulate as a level sheet in a basin. Four major categories of habitat are thus formed: (1) upper bajadas with coarse stony soils in which moisture sinks rapidly and deeply, (2) lower bajadas which have higher water retaining capacity but still have good internal drainage, (3) clay-floored basins (barrials or playas) on which water may stand for short periods and where the water-retaining capacity is highest but salts are usually excessive, and (4) floodplain–terrace systems.

Deflation during the long dry seasons, and water erosion during brief periods when water moves over the soil surface as a sheet, both leave a residue of pebbles and stones on the surface called "desert pavement." Sand-sized particles may be abundant enough locally to form dunes, but sandiness does not accurately characterize deserts. Even in the well-publicized Sahara Desert, sand covers scarcely more than 10% of the area.

Although the weathering of parent materials under desert conditions tends to yield coarse-textured products, some desert soils are largely clays, and most have clay in one or more horizons.[217] Where the texture is coarse, a fair proportion of the storms wet the soil profile deeper than 3 dm, and so provide moisture not susceptible to direct evaporation, therefore wholly available to plants. Below those horizons which are usually wetted, is a zone that remains dry, or nearly so, except following unusually heavy rainfall. Such events provide the occasions when additional seedlings of deep-rooted phreatophytes can make contact with permanent moisture in the capillary fringe above a water table. If there is no water table within reach, deep-rooted plants must use the deep, diffuse, and seldom replenished supply very sparingly during the long periods between recharges.

Since few rains penetrate deeply, soluble salts as well as carbonates produced by mineral weathering cannot be leached out of the solum readily, so saltiness increases with depth. These salts necessitate relatively low osmotic potentials (averaging -25 to -30 bars) in the sap of woody desert plants.[179] Also the soil reaction is slightly alkaline, which keeps some nutrients (Mn, Fe, Bo, Zn, Cu) in less than optimum availability.

Litter production is very low, and most is swept away by rain storms, so that humus is too scant to be evident to the eye even in the A horizon of desert soils. Thus, other than carbonate- or gypsum-cemented horizons, the soil profiles are weakly differentiated. The prevailing Red Desert Soils of North America are reddish owing to the oxidized state of iron that is not masked by humus.

By overflowing its banks annually, the Nile River formerly recharged its floodplain with soil moisture and at the same time added a thin film of silt replenishing its fertility. This unique situation provided a continuing supply of agricultural products for many centuries. Elsewhere agriculture in deserts has depended mainly on oases, water entrapment, or on water supplied by irrigation canals. Crops always need N, and usually P as well. Alkalinity keeps other nutrients in nearly unavailable form, and salinization is a major threat to continued agriculture. All irrigation waters, and especially those available to arid lands, keep bringing salts into the fields, and evapotranspiration leaves these behind to accumulate unless there is enough runoff to carry away at least as much as that coming in. Deep ditches are needed to carry the excess water away and prevent high water tables that would allow direct evaporation from the soil. Long ago the agriculture which developed just east of the Mediterranean Sea was based on the construction of small dams across drainageways, and ditches so arranged as collect runoff from large areas of desert and concentrate it on small gardens. The principle of salinization was not recognized, and many believe that the decline of early Mesopotamian civilization was a result of salt accumulation in the arable spots.

The *Larrea divaricata* Province

Rather surprisingly, the North American part of the Desert Region shares some 50 species or species-pairs with the deserts of temperate South America—*Acacia, Baccharis, Celtis, Cercidium, Cereus, Condalia, Encelia, Flourensia, Gutierrezia, Jatropha, Larrea, Lycium, Mimosa, Opuntia, Prosopis, Zizyphus.* These two segments of the Desert Region are now separated by a great gap, which suggests migrations of thousands of miles across the moist equatorial belt, or perhaps along rain shadows of the Andes that are no longer in existence.[376] There is some evidence that interglacials were hot enough to disrupt the continuity of tropical forests and so allow an exchange of xerophytes between north temperate and south temperate zones.[409] The major spread appears to have been from north to south, which may reflect the direction of sequential appearance of spots suitable environment.[330]

The coolness of Wisconsin time undoubtedly reduced the North American desert area well below its present size, perhaps limiting it to a small area centered on the mouth of the Colorado River.[281] Subsequent expansion from what may have been a single small refugium, probably explains the close floristic similarity of climatic climaxes throughout, and to be consistent with previous treatments, all the North American desert will be treated as a single Province. However, three Sections have long been recognized.

The Mojave Section[342,365] is a relatively small area situated in the angle between the San Bernardino and southern Sierra Nevada Mountains of California, extending eastward into southern Nevada, where climatic difference not clearly associated with mountains sets a limit. In this area desert has existed most if not all of the time since the mountainous arc around its western border uplifted at the close of Pliocene time.

Precipitation remains low, 50–125 mm a year, over a wide range of altitude from sea level to 1300 m above. Elsewhere in the *Larrea* Province there is a summer rainy season, but here rains are almost exclusively limited to the cold season—an influence of the Westerlies as they reach the southerly limit of their annual latitudinal shift (Fig. 13G). Perennials, and winter annuals that germinate at the start of the winter rains, mature quickly just before the Westerlies withdraw their influence in spring.

On the whole this desert lies a little higher in elevation and thus has lower minimal temperature in winters (to −12°C) than the other Sections. Understandably, the very few elements of the Artco–Tertiary Geoflora that have extended south as far as the Desert Region are almost all restricted to this Section.

Lower bajadas, comprising about ¾ of the area, are characterized by communities in which the evergreen shrub *Larrea divaricata* dominates, growing usually less than 1.5 m tall (Fig. 141). Smaller shrub associates include mainly *Franseria dumosa* and *Lycium*. The fact that *Larrea* retains its small varnished leaves (see Fig. 139) the year around attests to the existence of a continuous supply of available water within reach of its deepest roots. Climatic climaxes of this nature occur on Red Desert soils, and are open and monotonous owing to the high concentration of dominance in one relatively large plant. In southern Nevada, where the only quantitative studies have been made, shrubs, perennial forbs, and winter annuals each provide about 20% canopy coverage.[342]

Above an elevation of about 1000 m, especially in the central and eastern parts of the Mojave Section, is a richer mixture of shrubs hardly more than 1 m tall, these including *Acamptopappus sphaerocephalus, Aplopappus*

Fig. 141.   In all three Sections of the *Larrea tridentata* Providence, *Larrea* occurs in nearly pure stands with individuals widely spaced.

linearifolius interior, Eriogonum fasciculatum polifolium, Larrea divaricata, Lycium, Salazaria mexicana, and *Tetradymia.* Commonly the physiognomy is greatly enlivened by a scattering of tall *Yucca.* Most conspicuous among these is *Yucca brevifolia,* an endemic few-to-many-branched plant 2–10 m tall (Fig. 142). *Yucca schidigera* is not as tall or as branched, but is locally conspicuous. Stem succulents are sparingly represented in this Section of the Province.

Poorly drained soils, typically those peripheral to playas where the surface is dry except for a short period following heavy storms, support such halophytes as *Allenrolfea occidentalis, Atriplex canescens, Distichlis stricta, Sarcobatus vermiculatus,* and *Suaeda torreyana* (Fig. 143). Other special soil types support communities variously dominated by *Artemisia arbuscula, Coleogyne ramosissima, Grayia spinosa,* and *Lycium pallidum. Chilopsis linearis, Pluchea sericea,* and *Prosopis juliflora* are phreatophytes that line drainageways.

In the very low basin of Death Valley, which extends down to 480 m below sea level, there is very little vegetation except where surface water collects.

The larger native animals include mule deer, coyote, bobcat, badger, and kit fox. Feral burros are increasing, and if unchecked they will have disastrous effects on plant life, native animals, and soil.

Agriculture is nonexistent, and livestock grazing is very limited.

A more extensive desert area is centered on Sonora,[366] occupying almost the southern half of Arizona, much of Baja California, and in California extending up to the southern foothills of the San Bernardino Mountains.

Fig. 142. The bizarre growth form of *Yucca brevifolia* dominates the physiognomy of upper bajada vegetation in the Mojave Section of the *Larrea* Province.

This, the Sonoran Section, has about the same total precipitation as the Mojave Section, but it may come in either or both of two rainy seasons—one a season of gentle and protracted showers from December to March, a weak influence of the Westerlies, the other a period of convectional thunderstorms from July to September, a continental influence. Soils are wet more deeply by the winter rains. Although snow may occasionally fall in the

Fig. 143. *Atriplex* on a shallow, clay-floored basin in the Mojave Section of the *Larrea* Province. Although the species of *Atriplex* vary from place to place, the genus is characteristic of this land form throughout most of the Province.

higher parts, the winters are not so cold as in the Mojave Section (Fig. 13H). Eight to 12 months are frost free, and when freezing weather occurs it seldom lasts more than overnight.

Both floristically and morphologically the plants in this Section are more diversified than in the other two Sections (Fig. 144). The coarse soils of the foothills and upper bajadas bear a sprinkling of low trees (*Bursera, Cercidium microphyllum, Jatropha, Olneya tesota, Ipomoea arborescens, Parkinsonia, Pithecolobium sonorae, Prosopis juliflora,* etc.) that coalesce as gallery strips along intermittent stream courses. Among these scattered trees is a rich assortment of shrubs: *Acacia constricta, Condalia spathulata, Encelia farinosa, Franseria, Fouquieria splendens, Larrea divaricata, Lycium parishii, Simmondsia chinensis.* A third major component consists of stem-succulent cacti (*Cereus, Echinocereus, Ferrocactus, Lemaireocereus thurberi, Lophocereus shottii, Opuntia, Pachycereus,*) joined by leaf-succulents (*Agave, Dudleya*) (Fig. 145). In places these succulents and woody plants cover most of the ground, although they cast but little shade. Usually there is one or two species of cacti of the tall candelabra type, or a *Yucca,* that rises well above the stratum of low trees. It is these emergents, plus the abundance of low tree species, that give character to the Sonoran Section.

Perennial forbs are relatively inconspicuous, but several hundred species of small annuals make up for that floristic deficit. Both the forbs and annuals are divided into two groups that appear during just one of the two rainy

Fig. 144. Vegetation of coarse, stony soil near Florence Junction in the Sonoran Section of the *Larrea* Province. The tallest cactus is *Cereus giganteus;* the many-branched cactus at the left edge is *Opuntia fulgida;* the caespitose clusters of straight unbranched stems belong to *Fouquieria splendens.*

Fig. 145. *Agave deserti* with *Larrea divaricata* behind. Sonoran Section, California.

seasons (Fig. 13H). Those that respond to summer rains have floristic affinities with the continental interior. Those active in winter have affinities with southern California. The tiny short-lived *Tridens pulchellus* is the chief exception to the rarity of perennial grasses on zonal soils.

A little farther down the bajadas where the soils are more free of stones and gravel, yet are well drained, dominance is concentrated in *Larrea divaricata* and the small *Franseria deltoides*, these providing the strong element of similarity that draws the North American deserts into a closely related series.

At still lower elevations there may occur aggraded valley floors surfaced with silt and clay brought there by the heaviest rains. If there is internal drainage salinity is minimal and *Hilaria mutica* is a characteristic dominant, but if the water is held in the basin until it evaporates, salts accumulate to favor the dominance of *Atriplex polycarpa* or other halophytes such as *Allenrolfea occidentalis, Sarcobatus vermiculatus, Sporobolus wrightii,* and *Suaeda torreyana*. The Salton Sea basin is so low and saline that it has an extremely sparse flora (Fig. 146).

Stream terraces of the major valleys characteristically supported a closed deciduous woodland of *Prosopis juliflora* or *P. pubescens*, which was quite wide in places before irrigated agriculture preempted the habitats. Where the water table is near the surface, the alien shrub *Tamarix pentandra* has formed dense thickets that are of much concern for their influence in lowering water tables in the Sonoran Section, as well as in adjacent areas where water is in short supply. *Populus fremontii, Salix,* and a few other deciduous trees such as *Fraxinus velutina* and *Platanus wrightii* of Arcto–Tertiary affinities occur on the margins of permanent or near-permanent streams. Mainly in southern California and adjacent Baja California, *Washingtonia*

Fig. 146. Desert landscape near the Salton Sea, California.

filifera occurs in oases (Fig. 147) where a spring keeps a limited amount of soil moist, or in canyons where runoff from even small showers is concentrated.[407] Elsewhere in the Sonoran Section palms in other genera occur in similar habitats. Sand is seldom encountered in this Section, but where it occurs *Hilaria rigida* is a conspicuous dominant.

Characteristic vertebrates include mule deer, white-tailed deer, big horn sheep, peccary, jackrabbit, kangaroo rat, jaguar, puma, bobcat, coyote, kit

Fig. 147. *Washingtonia filifera* oasis in Palm Canyon, California.

fox, coatimundi, many snakes and lizards, and that especially interesting bird the roadrunner.

There is sufficient irrigation water coming from nearby mountains to the north for the development of agriculture in much of this desert. Some representative crops are cotton, melons, citrus fruit, olives, and dates.

The northern and eastern borders of the Sonoran Section, where ecotones are formed with *Bouteloua gracilis* steppe, chaparral, or *Pinus cembroides* woodland, coincides rather well with an isotherm beyond which the daily maximum temperature may remain below freezing in winter. The southern border, where it meets the Microphyllous Woodland Region, is related to the southernmost extension of frost. Westward the Sonoran Section gives way to the Mojave Section as winter temperatures become more extreme and summer rains lose effectivity.

Along the eastern edge of the Sonoran Section the continental divide consists of a gradual swell in a plain covered with steppe of the *Bouteloua gracilis* Province and studded with mountains. This swell approximates a climatic division separating the Sonoran Section to the west from the Chihuahuan Section that lies mostly east of the grass-covered divide. Winter rains brought onto the continent by the Westerlies hardly extend this far east, so that the summer rainy season of the continental interior provides nearly all of the annual 7–500 mm of rain (Fig. 13I). Winters are colder than in the Sonoran Section, with freezing temperatures continuing up to 72 hours.

There is little difference in physiognomy between the Chihuahuan and Sonoran Section on the upper bajadas. Over most of the Chihuahuan Section the tree life form and tall types of cacti are of much less importance, whereas small cacti, leaf succulents (*Agave lechuguilla, A. falcata, Hechtia*), sword-leaved plants (*Dasylirion wheeleri, Yucca*), and *Fouquieria splendens* are more conspicuous (Figs. 148 and 149). Toward its southern terminus dendritic cacti (*Lophocereus, Myrtillocereus*) appear, and there may be a question as to whether another Section should be recognized there. The rubber-producing shrub *Parthenium argentatum* is native to this Section.

On lower bajadas *Larrea divaricata* is again a major dominant, but here it is mixed with shrubs of similar stature such as *Flourensia cernua, Prosopis juliflora*, or *Acacia constricta* (Fig. 150).

Hilaria mutica flats and saline basins with *Atriplex* or *Sporobolus wrightii* occur here. Rather extensive dune tracts in this Section are stabilized by shrubby ecotypes of *Prosopis juliflora*. As in the Sonoran Section, *Prosopis juliflora* forms woodlands on stream terraces where the water table is not deep.

From the southwestern corner of Arizona this Section extends eastward into western Texas, and southward into Chihuahua and adjacent states of Mexico.

Fig. 148. Chihuahuan desert south of the Pecos River in Texas, with *Agave lecheguila* in the foreground, *Fouquieria splendens* and leguminous shrubs in the background.

Fig. 149. *Dasylirion wheeleri*, near Carrizoso, New Mexico.

Fig. 150. *Larrea divaricata* mixed with *Flourensia cernua* and *Acacia constricta* west of Tombstone, Arizona.

There are small highly disjunct areas of desert in some of the dry valleys of southern Mexico, but these are not well enough known to consider at this time.

SOME SIGNIFICANT FEATURES OF TROPICAL CLIMATES

Astronomically "the tropics" refers to the area between the Tropics of Cancer and of Capricorn (23½° N and S latitudes, respectively), but from a botanical standpoint a better definition would be the area essentially free of frost on lowlands. These frost lines are very sinuous, dipping equatorward in the interior of large land masses, then swinging poleward as they approach sea coasts (Fig. 151). Between these two isotherms floristic affinities are overwhelming with equatorial vegetation despite the sinuosity. Climates between the two frost isotherms have other important attributes as well, and these had best be understood before considering their vegetation.

On the equator where daylength is approximately 12 hours throughout the year, mean monthly temperatures are likewise nearly constant, the warmest and coldest months commonly differing by only 2°C. However, diurnal temperature variation is several times greater. These conditions obtain at all altitudes, so it is only outside the tropical belt that summer and winter temperatures become differentiated sufficiently to determine seasons of vegetative activity and quiescence. Since only well within the tropics are

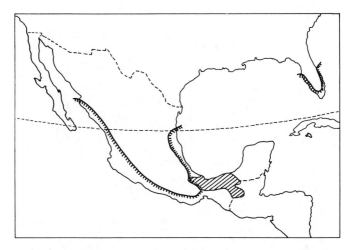

Fig. 151. Northernmost limits (ornamented line) of *Bursera simaruba* in Mexico and Florida, in relation to the Tropic of Cancer (dashed line). That portion of lowland rain forest which extends northward into Mexico is shaded. *Bursera* is a characteristic tree of tropical xerophytic forest in the Caribbean area, and is seral in rain forest.

monthly temperatures relatively constant the year around, only here does the mean annual temperature statistic have much ecologic significance.

Average temperatures decrease about 0.5°C for each rise in elevation of 100 m in the tropics. In Latin America it has long been common to recognize four altitudinal temperature belts, and Lauer[250] has suggested quantification of these belts as follows:

> *Tierra Helada*—mean annual temperature below 10°C, therefore treeless
> *Tierra Fria*—10–17°C
> *Tierra Templada*—17–22°C
> *Tierra Caliente*—above 22°C

Although the Spanish word *templada* is translatable as "temperate," the very small range of monthly temperatures here results in a climate and flora very different from that of latitudes beyond the tropics, and the term temperate will be used in this book for only those climates having warm summers and cold winters, with freezing weather during the latter. In the tropics frost is very rare below the alpine belt (*Tierra Helada*). However, on the mountains below, plants may be injured by prolonged exposure to temperatures several degrees above the freezing point of water.

Where temperatures are so nearly constant throughout the year the seasons are differentiated mostly by precipitation. Both the seasonal distribution and amount of precipitation are strongly related to latitude. In those months when the sun is overhead, net radiation (incoming solar energy minus the outgoing) is at maximum, and the greater heating of the earth's surface at this time tends to cause afternoon convectional showers in otherwise sunny days. At the equator the sun passes overhead twice a year, so the annual precipitation curve tends to have two equidistant peaks of convectional rainfall coinciding with the equinoxes. Progressing toward either tropic the two rainfall maxima come progressively closer together in an increasingly more pronounced warm season, until at the tropics there is only one brief season of zenithal (summer) rains. Owing to the humidity and high thin clouds of equatorial climates, the insolation received at the earth's surface is relatively low, its intensity increasing in either direction toward the Subtropical High Pressure belt where it is at maximum.

Just within the margin of the tropics rainfall is scant in addition to being confined to one short season in the warm half of the year. With little moisture to carry over into the long rainless season of cloudless skies, aridity is only a little less severe here than in the mainly extratropical Desert Region, and the Microphyllous Woodland Region stretches as a belt across the continent at this point.

Next in order toward the equator is the Savanna Region with a little more rainfall and a dry season that is not quite so long. Then on the equatorial side of the Savanna Region, where there are two closely spaced rainy seasons and the yield of water is higher, closed forest prevails, although the phenology of the vegetation is closely attuned to the alternation of wet and dry seasons. This is the Tropical Xerophytic Forest Region, with 5–9 months of rainy weather.

In the vicinity of the equator the two rainy seasons, equally spaced in the year, are so drawn out and involve so much moisture that drought stress during the short intervening periods is too feeble for vegetation phenology to be evidently affected. This is the ever-wet Tropical Mesophytic Forest Region, in which at least 10 months have abundant rainfall, and such rhythms in species phenology as exist are not drawn into a common pattern related to monthly variations in rainfall.

The local distribution of land and sea, mountains and lowlands, create frequent deviations from the idealized latitudinal pattern of rainfall and insolation described above, as would prevail everywhere if astronomic forces alone were governing. For example, throughout Middle America climates on the Caribbean coast are mostly of the ever-wet type, whereas the Pacific coast at the same latitude has a well-developed dry season in most places. And in Colombia and Ecuador surprisingly dry valleys are encircled by mountains on which rainfall is high.

The heat level in *Tierra Caliente* makes plants sensitive to even a short series of rainless days. Only heavy rainfall well distributed over the year can prevent drouth stress from developing under the high potential evapotranspiration that results from continually high temperatures and mostly sunny days. Thus the length of the dry season is the climatic parameter most closely

234

Part II. Ecologic Plant Geography

related to the distribution of vegetation types on the lowlands. (Compare Figs. 13I,J,K,L and 14A.) At high elevations low temperatures reduce the impact of rainless seasons.

Much less is known about tropical than about temperate and cold-climate forests, hence it is not feasible to attempt to divide the Regions into Provinces.

THE MICROPHYLLOUS WOODLAND REGION[24,363,364]

In the Microphyllous Woodland Region the climatic climaxes present a closed or nearly closed cover of microphyllous, mostly deciduous low trees and shrubs, which in the Americas frequently includes a scattering of columnar cacti (Fig. 152).

In North America this Region is bounded on the poleward side by the Desert Region, or by the *Bouteloua* Province. On the equatorward side it gives way to the Savanna or Tropical Xerophytic Forest Regions.

From the southern tip of Sonora it extends southward along the coastal lowlands of Sinaloa to Acapulco, with smaller portions at the tip of Baja California and in southern Mexico. A less extensive strip occurs on the east coast of Mexico. In the Caribbean Sea it is common on islands that have total relief too low to stimulate precipitation from the Westerlies, as well as on mountainous islands that have rain shadows of sufficient dryness. A large unit of microphyllous woodland locally called "caatinga" occurs in eastern Brazil opposite Recife,[258] and a narrow strip lies along the coast of Venezuela.

Fig. 152. Microphyllous (thorn) woodland beyond an artificial clearing, south of Mochis, Sinaloa, Mexico.

Broadly speaking, microphyllous woodland occurs in a latitudinal belt along the poleward margin of the Trade Winds, where moist tropical air brings only a very short period of rainy weather in summer. Then as the Trade Winds recede in winter, dry air descending in the subtropical high pressure belt moves equatorward over the earth's surface and the long rainless season begins (Fig. 13J).

Moisture is more plentiful in this Region than in deserts, chaparral, or steppe, but its effectivity is low since the showers are concentrated in a few months when the sun is overhead, and the trees stand mostly leafless during the remainder of the year. Where the rainy season is very brief rainfall may aggregate up to 1000 mm, but it is considerably less where showers are better spaced. In most of the Region, temperatures are rather uniformly high, but near the poleward limits the vegetation may be subject to light frosts.

The tallest stratum of the climatic climaxes is seldom as much as 10 m high, and is often considerably lower. Among the most widespread woody genera north of the equator in the Americas are *Acacia*, *Caesalpinia*, *Colubrina*, *Cordia*, *Ficus*, *Guiacum*, *Jatropha*, *Karwinskia*, *Lysiloma*, *Pithecellobium*, *Prosopis*, *Randia*, and *Zizyphus*. Columnar cacti of the arborescent stratum are provided by *Cephalocereus*, *Pachycereus*, and *Pterocereus*. Tall species of *Opuntia* may occur.

The stems of most of the nonsucculent trees are crooked and low-branched if not caespitose. In the Americas and in Africa the woody plants are mostly thorny and the vegetation is often referred to as "thorn woodland," but the Australian homolog is thornless. Their flowers are usually small and aggregated into conspicuous clusters.

The shrub layer which is approximately 1–4 m tall and commonly has higher coverage than the tree layer, may include the coarse terrestrial *Bromelia pinguin* or an occasional *Agave* or *Yucca*. Perennial and annual herbs may appear abundantly during the rainy season, but grass is essentially lacking. In places epiphytes, especially in *Tillandsia*, suggest abundant humidity, even though moisture is very limited for those plants which must get it from the soil. Slender-stemmed lianas may be common.

In Sinaloa permanent streams running through this Region are fringed by gallery forests containing *Acer*, *Carya illinoensis*, *Taxodium mucronatum*, *Salix*, and *Ulmus*. Broad-stream terraces may support nearly pure stands of *Prosopis juliflora*, with palms dominating the floodplains. Farther south moist ravines may contain topoedaphic climax forests representing extensions of the tropical xerophytic forest to be considered later. Elsewhere shallow soils have low open vegetation consisting largely of cacti and *Agave*.

For the most part microphyllous woodland is used for grazing or browsing, although it provides poor forage most of the year. Products harvested from the natural vegetation include lignum vitae wood (*Guiacum sanctum*), dye (*Haematoxylon campechianum*), special tannins (*Caesalpinia*), essential oils

(*Amyris*), and firewood. Some dry farming is possible in places, growing the sisal *Agave* or pineapple. Where irrigation water is available sugar cane, maize, banana, or tomato are grown.

THE TROPICAL SAVANNA REGION

The Tropical Savanna Region embraces all the tropical area in which the natural vegetation of zonal soils includes a layer of perennial grass, which is usually dotted with trees or shrubs growing either singly or in groves (Fig. 153). In rain shadows it may occur up to the alpine belt.

The word savanna by itself has become simply a physiognomic term indicating the presence of tall woody plants scattered over a matrix of lower vegetation, and so must be qualified to have any ecologic meaning. Even in the tropics there are three ecologically distinctive types of savanna.

In places there is sufficient moisture for trees or shrubs to maintain populations on zonal soils, but not enough for them to grow in sufficient density to exclude a substantial layer of xerophytic grasses. This is climatic savanna.

In wetter climates vegetation with savanna physiognomy may be the climax cover of areas where abnormal (infertile, shallow, or wet) soils prevent woody plants from forming a closed cover that would eliminate graminoids. This is edaphic savanna.

Finally, where tropical vegetation is destroyed and secondary vegetation is burned at frequent intervals thereafter, there are fire climaxes with

Fig. 153.   Savanna as climatic climax near Magdalena, Nayarit, Mexico.

savanna physiognomy which are appropriately called derived savanna. The following discussion will concern just climatic savanna.

To judge from the general pattern of the world's vegetation, a belt of climatic savanna would be expected on the equatorward side of the Microphyllous Woodland Region everywhere, with Tropical Xerophytic Forest on the opposite side. In this segment of the idealized climatic gradient the rainy season is slightly more protracted than in microphyllous woodland (compare Figs. 13J and K). Owing to aridity in tropical eastern Africa, on that continent the savanna belts of both hemispheres are connected. Isolated pockets of savanna also are to be found in deep tropical valleys where surrounded mountains supporting rainforest intercept all precipitation except during a few of the wettest months.

Although it is dry and is grass-dominated, tropical savanna soils lack the layer of lime accumulation that occurs in steppe.

Since the shoots of the grasses of climatic savanna die back to the ground during each dry season, this vegetation is highly susceptible to burning. Even without the intervention of man, lightning would at least occasionally start a conflagration here. Thus, fire is an integral part of savanna environment, and both the woody and herbaceous components have become adapted to this factor. Although fire is not essential for the maintenance of the physiognomy of climatic savanna, the species composition is always attuned to the frequency of burning.

Grazing that thins the grass cover can make more water available for woody plants and so may permit a thickening of their populations, which would, in turn, militate against grass reestablishment. Such vegetation resembles microphyllous woodland. Excessive grazing has in many places gone farther and eliminated woody vegetation as well as herbaceous plants, giving the land the aspect of a desert, especially on the poleward (drier) margin of the Region. Along the quatorward (wetter) side fire has converted a marginal strip of the adjoining forest to derived savanna. Thus superficially, the position of savanna as a physiognomically distinctive belt has been shifted equatorward, although the potentialities of the land have not changed permanently.

The limits of climatic savanna can be determined most reliably by the study of relics that have accidentally escaped conversion to fire or zootic climaxes, and exclosures such as may occasionally be found in hunting preserves. It has also been suggested that derived savannas may be recognized by their lack of endemics, for such vegetation is the relatively recent product of man's disturbance in most places, with the flora consisting of either fire-tolerant survivors of the former forest cover, or fire followers of wide geographic distribution.

Nineteenth century plant geographers did not appreciate the extent of derived savanna, which is understandable in view of the difficulty of evaluating the influence of primitive peoples in the semiarid belt. There is still much confusion and uncertainty regarding the status and boundaries of savanna as a climatic type in the tropics.

Tropical savanna trees usually have thick bark combining good insulation with low flammability. Should fire be hot enough to kill the shoots, they regenerate from adventitious buds near the summit of a thick tap root. Typically, they are single-stemmed, slow-growing, often gnarled and low in stature. Most are angiosperms, in some areas mainly deciduous, in others, e.g., Australia, mainly evergreen. Many are thorny. Sometimes the woody plants are only shrubs, and occasionally woody plants are entirely lacking[411] although such vegetation is still classed with savanna.

Savanna grasses are caespitose, narrow leaved, fairly deep rooted, and vary from short to as much as 3 m in height. The soil often does not dry to the depth of their root systems, as shown by the commonness of new growth appearing above ground in advance of the start of a new rainy season. Productivity is proportionate to rainfall (Fig. 154).

The grasses of all tropical savannas, whether climatic, edaphic or derived, have low nutritional value for grazing animals, in comparison with grasses of the Steppe Region. This is particularly true of derived savanna, where moisture supply and productivity are highest. Phosphorus and especially the protein content of the shoots are very low. Optimal protein content for cattle is about 10%, with approximately 5% being a critical minimum. In most tropical grasses protein rises barely above the minimum during their season of most vigorous growth, then commonly drops below 2% as the grass dries. In temperate zone steppe corresponding values are commonly 14 and 7%.

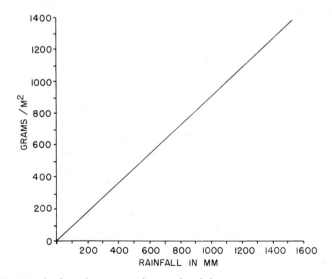

Fig. 154. Grassland productivity as dry weight of shoots per square meter per year, as related to annual rainfall in tropical seasonal climates.

An especially interesting feature of many savannas, whether climatic, edaphic, or derived, is the abundance of ground-dwelling termite colonies that build great moundlike nests, bringing soil particles of a particular size-class range up to the surface. In the course of time such activity results in a textural stratification that has a profound influence on plant life. In Africa there are savannas in which the trees occur as groves restricted to termite mounds. Depressions may accumulate salts here as in deserts.

Gallery forests in the Savanna Region have a rich and varied flora. In Mexico *Taxodium mucronatum* and *Salix* growing along streams are reminiscent of temperate zone vegetation.

That area of climatic savanna commonly called the Kalahari "desert," which lies just inland of the Namib Desert of southwestern Africa, is probably the most nearly pristine savanna area left in the world, despite a rapidly developing cattle industry. Its fauna of antelope, giraffe, ostrich, lion, and cheetah is exceptionally rich. Extensive savanna in central and northern Australia has a largely marsupial fauna. Central India also has another large area of savanna. Asian and American savannas lacked the rich megafauna so characteristic of Africa.

The chief land use in the Savanna Region is grazing. Agriculture is feasible where water is available for irrigation, but unlike grasslands of the temperate zones, the soils are not very fertile.

TROPICAL XEROPHYTIC (SEASONAL) FOREST REGION

The Tropical Xerophytic (Seasonal) Forest Region includes all areas in which the zonal vegetation is closed mesophyllous forest that is intolerant of frost and exhibits an evident seasonal rhythm in phenology that is attuned to alternating wet and dry seasons. The few grasses present in these forests are broad-leaved sciophytes.

Origin and Distribution

Seasonal forests developed as segregates from tropical mesophytic forest (rain forest), consisting of those elements which are preadapted to drought but intolerant of frost. The angiosperm component of temperate mesophytic forest, on the other hand, evolved as a very different segregate from tropical mesophytic forest, and consisted of those stocks preadapted to cold winters but intolerant of drought. Therefore, there is no taxonomic relationship between the forests of these two segregates, although there is often an astonishing physiognomic similarity between the deciduous components of each.

A close relationship between seasonal forest and rain forest is shown by their possession of some species in common, with ecotypes being deciduous in only the seasonal forest; by the frequency with which deciduous species play seral roles following deforestation in rain forest areas; and by the many rain forest species that extend into the Xerophytic Forest Region as phreatophytes (Fig. 155).

Seasonal forest occurs on all continents within the tropics. Large areas occur in Middle and South America, Africa, the Indo–Malayan area, and Australia. It tends to form a latitudinal belt which on the equatorial margin gives way to rain forest. At its poleward limits it is bounded by Microphyllous Woodland or Savanna Regions near the west sides of continents, and by the Temperate Mesophytic Forest Region on the east sides. Within the tropics it may extend up mountain slopes to essentially the bottom of the midmontane rainforest. In Mexico seasonal forest extends northward farthest as narrow strips along the seaward flanks of the Sierra Madre Occidentale as far as the 27th parallel, and along the Sierra Madre Oriental to southern Tamaulipas.[227] The southern tip of Florida, islands of the Caribbean, the north margin of South America, and the Pacific coast of Middle America in the same latitudes all have climates suitable for seasonal forest (Fig. 131). Close floristic relations throughout the Caribbean area reflect former land continuity, or at least much smaller water gaps separating Florida from Cuba

Fig. 155. *Sloanea quadrivalvis*, an evergreen tree common on uplands in rain forests of Middle America, is here restricted to the riparian habitat in semideciduous forest near Cañas, Costa Rica.

and Middle America. Apparently Florida was never directly connected with South America.[332]

Environment

The term seasonal forest came into use to distinguish this vegetation from rainforest where rainfall is so well distributed over the year that vegetation exhibits no seasonal aspection. Here in the Trade Wind belt the rainy season is longer than in the Savanna Region (compare Figs. 13K,L). The amount of rainfall is not as important as the length of the dry season. For example, the phenomenally high rainfall (11,610 mm) at Cherrapungi, India, is still insufficient for mesophytic forest, since four consecutive months have less than 100 mm. Considerably less rainfall will suffice for rain forest if it is better distributed through the year. It has been said that tropical climate should be considered seasonal if there are fewer than 40 rainy days in the 4 consecutive driest months.[403] Where the vegetation remains intact, the ecologist may prefer to base his judgment on its phenology.

At the dry margin of seasonal forest practically all the trees stand leafless during the rainless season. Progressing into areas with shorter dry seasons, more and more of the trees, especially the shorter ones, retain their leaves throughout these months, and the term semideciduous forest becomes appropriate (Figs. 156, 157, and 158). Finally, all but a few species in the top stratum are evergreen, yet their flowering is mostly concentrated in the short dry season, followed by a burst of leaf replacement. Whereas only two

Fig. 156.   Canopy of semideciduous forest near La Libertad, El Salvador.

Fig. 157.   Interior of semideciduous forest when in full leaf during the rainy season. Near Cañas, Costa Rica.

Fig. 158.   Interior of semideciduous forest in its most leafless condition. Same stand as in Fig. 157.

or three layers at most may be distinguished in deciduous forest, several are somtimes apparent in the evergreen vegetation.

Evergreen seasonal forest has almost as high a species diversity as rain forest; the trees are almost as tall and may be buttressed, and lianas and vascular epiphytes are usually well represented. Thus, in most physiognomic characters there is a close similarity to rain forest. However, the boles of the trees still tend to be crooked and low branched as elsewhere in seasonal forest, the canopy still has a relatively uniform surface, and there is a definite season of flowering and leaf renewal. This functional difference is considered the ecologically more sound and objective basis for dividing tropical forests, but others have emphasized leaf persistence and lumped all evergreen members of seasonal forest with rain forest. Where that is done, considerably more land would be included in the Rain Forest Region than is taken in here. Obviously at the wetter end of the seasonal catena the term "xerophytic," although always relative, is barely acceptable.

Nearly all the early literature on tropical xerophytic forest concerned the "monsoon forests" of southeastern Asia. There, air masses moving north off the Indian Ocean bring heavy rains during the warmer part of the year, then a reversal of air mass movement in the alternate season brings dry air descending from the desertic plateau of central Asia desiccating the lowlands as it flows toward the Indian Ocean. In contrast with the Americas, Asiatic seasonal forests commonly include communities dominated by tree-size bamboos.[13] In parts of that area *Dipterocarpus*, *Shorea*, and *Tectona* are widespread genera of commercial importance. Australia has large areas of seasonal forest in which mainly evergreen species of *Eucalyptus* are the major dominants. South America includes large tracts of similar forest, especially south of the equator.

Owing to the temperature-ameliorating influence of adjacent seas, coastal strips of seasonal forest extend north of the Tropic of Cancer in Mexico and around the southern tip of Florida. This forest may have reached Florida rather recently, for during the Hypsithermal Interval it appears that *Quercus* savanna occurred on the peninsula, with *Pinus* and additional angiosperm trees invading only as moisture became more plentiful.[415]

Progressing southward across the temperate mesophytic forests of the eastern United States precipitation increases, but becomes more and more restricted to summer. As the winters become drier, the trend toward evergreenness in woody vegetation eventually becomes reversed, until the southern tip of Florida falls clearly in the seasonal forest Region, with a large percentage of the trees deciduous in the dry months.[7-9] On the wetter east side of the tip of the peninsula evergreen species are better represented in the relics and offshore islands than they are on the drier west side.

The soils of seasonal forests are circumneutral and relatively fertile. In a semideciduous forest studied in Costa Rica it was found that the soil profile dried to the wilting point to a depth of 30 cm and remained this dry for three months.[105]

During the dry season surface fires frequently run through these forests. However, there is little to burn except the recently fallen leaves, so that well-established trees survive by virtue of bark thick enough to protect the cambium, and seedlings that are killed back to the soil surface sprout promptly.

Zonal Ecosystems

Seasonal forest differs from savanna in that the tree canopy is essentially closed, and the few grasses which occur beneath the canopy are broad-leaved sciophytes. It differs from microphyllous woodland in being mesophyll-ous, taller, and essentially without thorns in the dominant stratum, although in the drier parts of seasonal forest the shrub layer commonly has many thorny members.

At the dry margin of seasonal forest where the trees of the dominant stratum are all or mostly all deciduous, some shrubs and tree seedlings may retain their leaves throughout the dry season. The boles are relatively sturdy and low branched (Fig. 159). Buttresses are usually lacking. Slender aerial roots often descend from the boughs, eventually penetrating the soil. The taller trees have leaves mostly in the mesophyll class, but smaller leaves prevail among the shorter trees and shrubs. The leaves are membranous in the deciduous species, and when young they are protected by specialized bud scales.

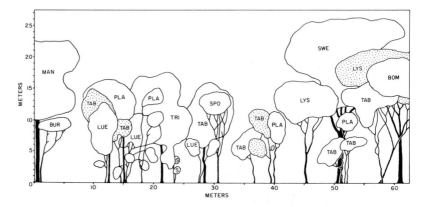

Fig. 159.   Profile diagram of a semideciduous forest in Guanacaste Province, Costa Rica, showing all trees 5 cm or more in diameter at breast height, as encountered on a strip of land 8 m wide: *Bombacopis quinata, Bursera simaruba, Luehea candida, Lysiloma* sp., *Manilkara spectabilis, Platymiscium dimorphandrum, Sapranthus palanga, Spondias purpurea, Sweetia panamensis, Tabebuia neochrysantha, Trichilia colimana.*

In contrast with nearly all other forests, the flowers of these trees tend to be large and conspicuous, appearing during the dry season when the deciduous trees stand leafless. Insects, birds, and bats are the pollinating agents. Cauliflory is uncommon (Fig. 160) and is largely associated with bat pollination. Although leaflessness seems logically associated with the dry season through moisture stress, it is a curious fact that a few species drop their leaves before the rainy season ends, and during the dry season most trees produce a profusion of large delicate flowers that would seem quite vulnerable to rapid water loss. Furthermore, the deciduous trees commonly put forth their new leaves before the start of the rains, and their seedlings are often evergreen for the first few years. The chief evidence that there is truly a water stress consists of marked shrinkage of tree trunks during the dry season,[106] and among the woody plants which leaf out prior to the first rains, the new leaves may hang in an evidently wilted condition until it rains. Flowering and precocious leafing are made possible by the moisture that is not completely exhausted below the depth of ½ m or so.

Bud burst is most likely triggered by daylength, for it has been demonstrated that even plants near the equator are sensitive to small variations. The appearance of flowers during the dry season probably developed primarily as an adaptation to insect pollination, for insects are most active during this part of the year, and the lack of foliage not only makes the plants more visible from a distance, but it interferes less with insect flight.

Palms and bamboos are often represented in seasonal forest, with lianas, geophytes, and annuals common. On the other hand, conifers ferns, and

Fig. 160.   Cauliflorous trees like this *Crescentia alata* bear flowers and fruits from buds on old wood, rather than on young twigs as in other trees.

mosses are scarce. Vascular epiphytes too are usually few (see Fig. 155). Some representative tree genera of the deciduous and semideciduous forests bordering the Caribbean Sea are *Bursera, Cecropia, Ceiba, Coccoloba, Cochlospermum, Cupania, Enterolobium, Guazuma, Hura, Hymenaea, Pisonia, Pterocarpus, Sideroxylon, Spondias, Swietenia, Tabebuia, Trema, Trichilia,* and *Zanthoxylum.*

Derived Savanna

Vegetation in which the foliage is abscised during a hot dry season (Fig. 161) naturally becomes highly flammable, so fire has long been an important environmental factor in seasonal forest, roughly in proportion to the percentage of deciduous trees. As a result of man's aggravation of this hazard, most of the deciduous and much of the evergreen seasonal forest has been converted to fire-maintained savanna the world over (Fig. 162).[43] Where primitive agriculture is practiced, land deforested for cropping becomes progressively infested with weedy grasses difficult to control with a hoe, and it is then abandoned in favor of a newly made forest clearing. Forest cannot invade this grass cover of an abandoned field if it is frequently burned, so savanna is maintained as a fire climax. In Africa elephants devastate trees by stripping their bark for food, and this is an additional factor maintaining the vast arc of savanna derived from seasonal forest that surrounds the Congo rain forest.

Fig. 161. Recently cast litter in a tropical semideciduous forest has been redistributed by strong winds during the dry season, so that slightly convex surfaces are swept bare. Same stand as in Figs. 157–159.

Fig. 162. Fire-maintained savanna dominated by *Hyparrhenia rufa*, that has replaced semideciduous forest near Cañas, Costa Rica.

The trees in derived savanna may represent species residual from the original forest that were preadapted to a burning regime, or they may be fire-tolerant species that have invaded from adjacent climatic or edaphic savannas. Trees from the last two sources may not be able to reach small fire-maintained clearings that are maintained deep in a forested area, and in fact, the tree element may not be present at all if there are no fire-adapted species available. In the Americas common trees encountered in derived savanna include *Acacia, Bowdichia, Byrsonima, Cassia, Curatella, Erythrina, Inga, Mimosa, Pinus,* and several genera of palms. The narrow-leaved heliophytic grasses in these communities may include natives that have spread from climatic or edaphic savannas, but for the most part they represent fire-tolerant introductions from the Old World tropics such as *Hyparrhenia, Imperata, Panicum maximum,* and *Pennisetum.* Once the tree cover is thinned sufficiently for these grasses to form an understory they provide much fuel so that fires are then hot enough to devastate the seedlings of woody plants unless they are especially adapted to endure fire.

Southern Florida uplands subject to frequent burning develop into fire climaxes dominated by *Pinus elliottii densa,* these at first having an undergrowth that includes roots sprouts of the dicotyledonous trees that were formerly the dominants. With continued burning the canopy opens leaving a few shrubby palms or grasses, e.g., *Andropogon, Muhlenbergia, Sorghastrum* etc. to form the savanna ground stratum. Similar fire-induced and maintained savannas with *Pinus caribaea* occur in Cuba, and at the west end of the Caribbean Sea.

To prevent the encroachment of woody plants leading back toward climax forest, savannas must either be burned regularly (Fig. 163), or mowed. But fire during the dry season leaves the soil exposed to wind erosion, and then when the rains begin the nearly bare soil is vulnerable to water erosion.[105] Mowing as a management technique for grazing lands largely prevents serious erosion and makes it possible to maintain species of higher nutritional value (e.g., *Digitaria procumbens*) than those tolerant of fire, but in most areas mowing is either impractical or uneconomical, hence degradation of derived savanna is almost universal. There is considerable apprehension that savannazation can have other undesirable influences on environment in addition to allowing serious erosion. The soils dry more deeply than under forest conditions,[105] and the water table may decline. Reduced transpiration returns less water vapor to the air, and since much rainfall over the land is derived from evapotranspiration, it has been reasoned that savannazation may indirectly reduce rainfall.[15] Some have hypothesized that large areas where savanna occurs on shallow soil over a hardpan once supported forest that was destroyed by man, and the hardpan represents irreversible soil deterioration.[57] These hypotheses are difficult either to refute or defend, but there are grasslands in Ceylon which the surrounding native trees will not invade even where there is protection from fire, unless an alien *Eucalyptus* stand is first established to ameliorate the environment.[195] Such savannas at least border a condition of irreversible degradation of forest soil. And there are recent instances of soils becoming ruined by deforestation when laterite

Fig. 163. Maintenance of savanna derived from semideciduous forest requires either frequent burning or mowing. A back-fire advancing toward the camera is killing the root sprouts that developed since last year's fire. Costa Rica.

dries, although this does not necessarily mean that all savannas with hardpans are anthropogenic.

There is relatively little seasonal forest left for study, and it is very difficult to draw a line separating savanna derived from such forest from climatic savanna of the contiguous Region. Rain forest in the lower Congo Valley is surrounded by a vast arc of land that has been converted to savanna by man through burning practices that started before recorded history.

Nonzonal Vegetation

Mangrove is a major category of halophytic vegetation common to tropical and subtropical coastlines with fine-texture soils.[48] It consists of sclerophyll-ous evergreen scrub (especially where hurricanes are frequent), woodland, or forest. It extends from the shore into the shallow water of coastlines and deltas subject to negligible wave action, into estuaries, and up rivers a little beyond tidal influence.[69,263]

Mangrove is essentially a one-layered community, since undershrubs, herbs, and lianas are rare, and epiphytes may occur in only the margin most remote from salt spray influence. The woody members are distributed among 8 families and 12 genera. Southeastern Asia has 36 species, but only 10 occur in the Americas. Although poor in total species, most species are exclusive to it.

Each segment of the environmental gradient from sea to land tends to be dominated by a single mangrove species, and where there is a moderate to steep gradient, distinctive belts may be recognized. Since each species has its own range of tolerance of water depth or duration of tidal flooding, the belts may slowly advance or recede in response to changes in sea level or land elevation. In places they promote the deposition of mud or peat, and so tend to advance upon the water. In addition to being saline or brackish, the substrate is anaerobic.[375]

In Florida four zones are involved: *Rhizophora mangle* (next the open water, Figs. 164 and 165), *Avicennia nitida*, *Laguncularia racemosa*, and *Conocarpus erectus* (farthest inland).[110] Along the Pacific coast of the Americas, *Pelliciera rhizophorae* is another important mangrove tree. Along the inland borders of mangrove swamps where there is minimal salinity the pantropical shrub *Hibiscus tiliaceus* or the giant fern *Acrostichum aureum* are commonly encountered. In about the same position in eastern Asia the stemless palm *Nipa fruticans* grows in abundance.

Rhizophora mangle is especially renowned for its well-developed tangle of adventitious roots that arch outward from the trunk then branch profusely after penetrating the mud, thus supporting the plant well in its soft sub-stratum. These prop roots, commonly ensheathed in algae, barnicles, shellfish, or other invertebrates, are efficient conductors of gasses between the free air and the submerged parts of the root system. A number of

Fig. 164. Advancing edge of a *Rhizophora mangle* thicket, Florida.

Fig. 165. Rhizophora mangle, often a shrub only a few meters tall, can grow to tree size, as here on the margin of White Water Bay, Florida.

mangrove species are viviparous, that is, their seeds germinate while the fruits remain attached to the plant. In *Rhizophora mangle* the radicles extrude about 2 dm before the fruit drops, and in this condition the seedling floats and is disseminated by water currents. The disseminules of other mangrove species too are buoyant and tolerant of long submergence in sea water, so the species are widespread.

Approximately 75% of the world's coastlines between 25°N and 25°S latitudes support mangrove, regardless of the character of the vegetation on adjacent uplands. Owing to the temperature-ameliorating influence of the adjacent seas they extend farther into extratropical latitudes than does seasonal forest—to 32° on Bermuda. Dry climate is no limitation, for along the west coast of Mexico mangrove extends north until opposite the Microphyllous Woodland Region.

Mangrove on both sides of the Americas and western Africa share mostly the same species, whereas a different mangrove flora extends from the east coast of Africa to southeastern Asia. This floristic patterns results from (1) the absence of effective migratory barriers between east Africa and the East Indes, (2) the effectiveness of frost at the Cape of Good Hope as a barrier to migration around the southern tip of Africa, (3) the success of relatively short-distance trans-Atlantic dissemination in contrast with the failure of trans-Pacific crossing, and (4) a former sea connection across the Isthmus of Panama.

Like salt marshes and estuarine ecosystems, mangrove plays an important part in the life cycles of many species of aquatic animals, including fish and shrimp, since it shelters the young of some species, and its detritus washing into shallow water serves as the base of food chains for others. The submerged prop roots and stems often provide anchorage for invertebrates and algae which otherwise would be absent, and the canopies provide nesting sites for numerous colonies of waterfowl.

Those tropical shorelines which are subject to vigorous wave action and so have sandy beaches (if not rock-bound), support vegetation very different from mangrove. Whether the climate is seasonal or ever-wet, the strand zone just above the usual reach of waves is typically dominated by trailing evergreen plants such as *Ipomoea pes-caprae* and/or *Canavalia maritima* (Fig. 166). Frequently in the Caribbean area one or more erect plants such as *Distichlis spicata, Scaveola keonigii, Sesuvium portlaccastrum, Sporobolus virginicus,* or *Uniola paniculata* occur scattered among or in place of the trailing shrubs.

Inland from this low community, on sand that is slightly more elevated, are a series of salt-tolerant shrubs and low trees, some evergreen and some deciduous, which in the Americas include *Chrysobalanus icaco, Coccoloba uvifera, Colubrina, Croton, Eugenia, Erythroxylon, Hippmane mancinella, Terminalia cattapa* (introduced from Asia), and *Tournefortia gnaphalodes.* In Asia *Barringtonia, Casuarina,* and *Pandanus* are characteristic representa-

Fig. 166. This grass, *Uniola paniculata*, growing about 2 m tall, and the prostrate *Ipomoea pes-caprae*, are widespread species of tropical coasts where sandiness and salt spray are major factors of the environment.

tives of this belt. The diminishing influence of salt spray inland usually results in a segregation of these dominants into a series of belts, with dwarfed and misshapen individuals of the most hardy species next to the beach zone. These plants, as well as those of the beach, are mostly adapted for salt water dissemination and so are widespread. The communities are usually one layered, lacking epiphytes as well as herbs. Much of the area formerly occupied by them has been converted into plantations of *Cocos nucifera*, although palms are rare in the native vegetation of the belt, as are bamboos and ferns.

Over wide area in Middle and South America seasonal forest is interrupted by savannas, edaphic climaxes coextensive with thin soil over bedrock, in which two low trees, *Curatella americana* and *Byrsonima crassifolia*, are characteristic dominants (Fig. 167). The native grasses of the continuous ground layer vary greatly in composition from place to place. *Curatella* and especially *Byrsonima* may spread onto zonal soils after the seasonal forest has been destroyed and then prevented from returning by repeated burning, for both species are highly fire-resistant, and can sprout readily from roots should the trunk be killed. In northern South America *Bowdichia virgiloides* is an additional codominant in these edaphic savannas.

In addition to shallow soils, edaphic savannas often are associated with infertile parent materials, or with hardpans that result in soil saturation during the rainy season, followed by dry-season desiccation of the restricted root space. Along the Caribbean coast of Honduras *Pinus caribaea* joins *Curatella* and *Byrsonima* to dominate infertile sands that are alternately flooded and desiccated with the changing seasons. Savannas dominated by *Crescentia alata*, *Acacia farnesiana*, or by palms occupy other variants of the

Fig. 167. Savanna with *Curatella americana* and *Byrsonima crassifolia* as woody dominants is here an edaphic climax on a pumice outcrop, with semideciduous forest on adjacent land with deeper soil. Costa Rica.

alternately wet and dry habitats (Fig. 168).[28] South of the Amazon River in Brazil is a large area of edaphic savanna (the "campo cerrado") with deciduous or evergreen forest occupying the limited areas of zonal soils.[124]

Grumusols, areas of the black montmorillonite clay, which are excessively wet in the rainy season then dry and crack deeply in the alternate season, are common, and these support still different kinds of savannas.

Fig. 168. Savanna dominated by *Acacia farnesiana, Calliandra* and other low trees coincides with a flat lava flow which drains very slowly and is thus excessively wet during the rainy season, then dries severely in the dry months. Near Libertad, Costa Rica.

In tropical America river margins and freshwater swamps, especially those just landward from coastal strips of mangrove, often support a complete cover of palms (*Acoelorraphae, Bactris, Mauritius, Orbigyna, Raphia,* or *Scheelia*) which may be overtopped by scattered evergreen dicot trees. Elsewhere, freshwater swamps may be dominated by trees in *Bravisia, Ficus* or *Pterocarpus*. Typically the muddy surface of swamp soils is inundated for so long during the rainy season that almost no shrubs or herbs can survive there. Usually only one or two species dominate the canopy.

An extensive fen, bog, and marsh complex, the "everglades," covers over a million hectares of seasonally flooded land in southern Florida. It is dominated by *Mariscus jamaicensis* projecting 2–3 m above the water (Figs. 169 and 170), with scattered elevated spots supporting trees or shrubs.[123,262] Associated swamps have *Taxodium distichum* or the aliens *Casuarina equisetifolia, Melaleuca quinquenervia,* or *Schinus terebinthifolius*. Artificial drainage, which started in the 1930s, has lowered water tables, which rest on corraline limestone, until fires have eroded deeply into the peat and altered species balances.

Land Use

Although virtually no silviculture is practiced in seasonal forests of the American tropics, they are a rich source of valuable timber trees, including *Chlorophora tinctora* (fustic), *Cedrela* (Spanish cedar), *Cordia alliodora* (laurel), *Dalbergia* (rosewood), *Swietenia* (mahogany), and *Tabebuia*. As yet these trees are simply exploited, with little attempt to select superior strains and propagate them. In southeastern Asia seasonal forests contain two

Fig. 169. *Mariscus jamaicensis* fen with *Sabal palmetto* conspicuous on small elevated places in the Florida Everglades in June of 1936. At this season the water level had risen about 1 m above the low level of the preceding dry season, at which time no more than about half the Everglades was submerged.

Fig. 170.   Dwarfed *Taxodium distichum* and *Mariscus jamaicensis* growing on very thin soil over limestone in Everglades National Park, as seen in 1954. By this date extensive drainage canals had been dug, lowering the water levels so that much less land was inundated, and fires had eroded deeply into the accumulated peat.

much-exploited timber trees, *Tectona grandis* (teak) and *Shorea robusta* (sal), both of which are fire-tolerant and seral. The former has been introduced to a limited extent in Middle America.

Of all the zonal soils of the tropics those of seasonal forest are the most amenable to cultivation. They are circumneutral, have a high cation exchange capacity, and so are relatively fertile for these latitudes, especially in the deciduous and semideciduous types of forest. Furthermore, the alternation of strikingly contrasted wet and dry seasons periodically reduces insect and fungus populations, whose numbers in the wetter tropics tend to build up progressively to intolerable levels. Representative crops include cassava, citrus fruits, cotton, grain sorghum, maize, millet, peanuts, sesame, sisal, and upland rice. Paddy rice is grown where irrigation water is available, with sugar cane and banana very productive where the water table is not far below the ground surface. Cultivation is necessary during the wet season when the moisture content of the soil is usually supraoptimal for such manipulation.

Savanna derived from seasonal forest is widely used as cattle pasture. Although the productivity of the grass is high, the herbage is of low quality, being low in protein and nutritive minerals and high in silica. Since the protein content is highest when the grass shoots are young, some derived savannas are managed by burning portions at staggered intervals so as to maintain a continuing supply of fair quality forage. Despite the low quality of the grass it is extensively used for beef production, using the well-adapted

Fig. 171.   Overpopulation is leading to disastrous erosion as xerophytic forest land unsuited to cultivation is cleared for subsistence farming. South of Tuxtla Gutierrez, Mexico.

zebu. Competent ecologists have pointed out that in African savannas the use of the whole complement of native herbivores is the most efficient means of producing meat on a sustained basis, but their advice has gone unheeded by cattlemen. Owing to intense rainfall during the wet season, erosional losses can be severe either on slopes where savanna is maintained by annual burning, or where forest has been cleared for cropping (Fig. 171).

Mangrove is widely used as a source of fuel or charcoal, with the larger trees providing timber. The bark of some species is high in tannin, and the dye catechu comes from *Ceriops* of the Asian mangroves.

THE TROPICAL MESOPHYTIC FOREST (RAIN FOREST) REGION[26,318,359,385]

Included here are areas where evergreen forests occur on zonal soils in frost-free climates, where seasonal variations in moisture stress are so feeble that the vegetation as a whole shows no synchronous rhythms in phenology that are related to drought. In Java 68–79% of the plants in such forest are in flower every month of the year.

Origin and Distribution

This Region lies astride the equator, extending outward from centers of wetness until drouth becomes of sufficient importance to cause seasonal synchronization of activity. This will be taken as a reasonably objective boundary between mesophytic and xerophytic vegetation in the tropics.

One large but discontinuous segment extends from southern Mexico to Brazil. A second segment occurs in the lower Congo drainage and Guinea in western Africa. The third unit consists of tracts scattered over India, Malaysia, and Australia.

These three major areas containing tropical mesophytic forest, often referred to as rain forest, are not closely related floristically, for although their floras are very old, in tropical latitudes oceanic barriers to intercontinental migration became quite formidable soon after angiosperms began spreading over the land. Even so, there are a number of pantropical genera, and even some pantropical species, e.g., *Gyrocarpus americanus*.

Where the disseminules are heavy, as in *Gyrocarpus*, these floristic similarities must reflect species resisting evolutionary change that gained widespread distribution before continental drift created the barriers.[17] Other floristic relations may represent rare long-distance dissemination.

Fossil deposits in Panama,[385] Trinidad,[31] Borneo,[396] and the Indo–Malaysian area,[302] have all suggested continued occupancy of these areas by rain forest since at least early Cenozoic time—a situation without parallel in other types of the earth's vegetation.

During the Eocene Epoch rain forest expanded poleward until in the north it occupied about the southern third of North America, with a peninsular extension next the Pacific Ocean as far as southern Alaska. Mangroves, lianas, palms, cycads, and figs were among the tropical elements in southern Alaska, but virtually throughout the extent of this forest there was an admixture of genera which today are characteristic of only temperate climates.

Later, as climates cooled, rain forest retreated, withdrawing south of the Tropic of Cancer as glaciers accumulated. During interglacial periods, however, it must have spread northward almost as far as it had in Eocene time, for the remains of tapirs, which are strictly tropical today, have been found as far north as Ohio and Indiana. There is some evidence that during interglacials, when temperatures were higher than at present, rain forest in South America may have become fragmented into disjunct areas separated by dry tracts.[286,376] During at least the last southward retreat, some preadapted species of tropical affinity lagged behind to become integral parts of the temperate mesophytic forest which replaced it in southeastern North America: *Diospyros virginiana* (*Ebeneaceae*), *Catalpa* and *Tecoma* (*Bignoniaceae*), *Asimina* (*Anonaceae*), and *Sassafras albidum* (*Lauraceae*).

Environment

At low altitudes mean annual temperatures in rain forest may be as high as 28°C (Fig. 14A). Despite uninterupted physiologic activity the year around, primary productivity is not much higher in this Region than in the Temperate Mesophytic Forest Region (Table 3), presumably because respiration remains high during night as well as day.

Annual rainfall can be as low as 1700 mm if it is well distributed over the year, but monthly values seldom drop below 100 mm, which, owing to the abundant solar energy, could prove critical if two or more consecutive months dropped this low. The entire soil profile under forest cover remains moist throughout the year, as implied by the term "ever-wet" climate.

Since most rainfall on lowlands is convectional, the rains tend to be heavy but of short duration, and occur most frequently in afternoons. Often they are accompanied by lightning, and usually occur in 2 out of every 3 days.

Humidity is high, usually above 80% during the day and rising to essentially 100% at night when dew commonly accumulates on foliage. Although relative humidity may rarely drop below 50%, under the prevailing warmth the saturation deficit, i.e., the evaporative power of the air, is quite high. In terms of human comfort the climate is usually described as "muggy." Owing to this high moisture content of the air and the prevalence of high thin clouds, insolation is less intense here than in the subtropical belt, yet with the brevity of the showers most daylight hours are sunny. The precipitation/evaporation ratio of the equatorial belt approximates that of temperate climates, but this average includes much more than just rain forest tracts (see Table 2).

The term rain forest has often been applied to vegetation outside the Tropical Mesophytic Forest Region as defined here, especially to the ever-green phase of seasonal forest. It is also used locally in connection with a small part of the *Tsuga heterophylla* Province on the windward side of the Olympic Mountains in Washington, where the approach effect increases precipitation to about 3560 mm a year. However, the winters there regularly have considerable freezing weather with frequent deep snows, and the flora, fauna, and soils all differ sharply from anything within the tropics. The term rain forest has even been used for places in the *Picea engelmannii* Province where frost is frequent even in the warmest month.[211] By using the term "tropical rain forest" (A. F. W. Schimper's term), or "selva" (Latin American term) or "hylaea" (A. von Humboldt's term), such objectionable usages might be excluded. For the sake of brevity the term rain forest alone will be used here for all nonseasonal forests in the tropics.

Ever-wet climates of the tropics may extend from sea level up to timberlines at 335–366 m. This embraces a mean annual temperature range of about 20–10°C, and involves differences in fog and insolation as well. Thus, rain forest can be subdivided into altitudinal belts, and the following will be recognized here: lowland, lower montane, midmontane, and upper montane rain forest.

Lowland Rain Forest[312,337]

Lowland rain forest (lowland here referring to altitude above sea level, rather than connoting phreatophytic environment) ranges from sea level to about 1000 (250–1500) m above.

The weathering of parent materials procedes rapidly here, for temperatures are uniformly warm, moisture plentiful, and the lush vegetation releases acids as its litter decays. In this chemical environment Fe and Al remain insoluble, whereas silica is converted to silicic acid and is leached away along with basic ions. The red to pink clay (kaolin) that is formed in this process consists chiefly of residual Fe_2O_3 and Al_2O_3 that have two peculiarities of special biologic importance. With exchange capacity less than 15%, they have low capacity to hold the adsorbed basic cations needed by plants. Therefore, these cations are eventually lost and the soils become moderately acid (pH 4.5–5.0). Secondly, the residual oxides confer a porous, nonsticky property to the profile that allows rapid percolation, hence, rapid loss of ions that are not adsorbed. The effect of both these properties in promoting infertility is then enhanced by the rapid decay and mineralization of litter, which keeps the humus content at a low level. Although leaves are dropping almost continually, they decay so rapidly that the soil retains barely enough cover to exclude epigeous mosses. The processes described above are collectively referred to as laterization. The end product, laterite,[370] hardens irreversibly to bricklike texture if it is dried, and it has indeed been shaped then sun dried and used for buildings that have endured for centuries. Even simple deforestation will allow the soil to dry sufficiently for laterite to harden in 1–5 years, making agriculture impossible and retarding, if not preventing, the return of forest. Large areas of hardened laterite supporting savanna are believed to have been caused either by cropping in prehistoric times, or by a reduction in rainfall after an area had long supported rain forest. Since man has been in the New World for only about 10,000 years, the first of these possibilities seems unlikely, but the Hypsithermal Interval may well have created enough drought locally to have hardened laterite. Not long ago a large resettlement project in Brazil was abandoned only a few years after establishment, owing to rapid soil deterioration, and now a new highway ("Transamazonica") has been started from the Atlantic coast on the east, to the Peruvian border on the west, to open up a vast area of rain forest across north-central Brazil. Brazilian ecologists are understandably pessimistic regarding this project.

Lush vegetation persisting on soil of such low fertility seems a paradox. However, a large share of the nutrient capital of a rain forest ecosystem is held in the living organisms, and with the decay of each unit of organic debris proceeding rapidly to completion, coupled with intense competition for the nutrients as they are released, nutrient cycling rapidly brings ions back into plant tissues. About 50% of the fine rootlets are in the top decimeter of the soil, and mycorhizal hyphae have been reported as emerging above the mineral soil to penetrate fragments of litter. Thus, there may possibly be a short cut in cycling, with nutrients being relayed from one generation to the next without ever having entered into the exchange complex of the soil, or the soil solution. Owing to efficient cycling, nutrient losses from a mature soil profile are so low as to be compensated by rainfall input, and most streams draining rain forest are remarkably free of solutes except during flood stage. Although rapid growth in these forests suggests evolutionary adaptation to take up nutrients as fast as they are released by decay, the nutritive value of plant tissues per unit volume is relatively low.

In areas where occasional showers of volcanic ash add fresh supplies of primary minerals to the surface, soil fertility is maintained at a high level. Also on deltas and floodplains that are inundated during the rainy seasons (e.g., along the Amazon River), the fresh increments of silt maintain high fertility. Elsewhere, ancient soils subject to laterization become bodies of nearly pure Fe and Al hydroxides (up to 20 m deep!) and are valuable sources of hematite and bauxite, respectively. TiO_2 also accumulates to commercially valuable proportions in some of these profiles.

Three-quarters of all angiosperm families display maximum species diversity in the tropics,[19] and in lowland rain forest species diversity is higher than in any other type of forest, this being a consequence of large numbers of genera rather than many species per genus. Lowland rainforest is also unique for the fact that there are many more species of trees and shrubs than of herbs, approximately 70% of the flora falling in Raunkiaer's mega-, meso-, and microphanerophyte classes. In the Amazon Valley the ratio is 7:1, whereas progressing northward the ratio diminishes, then becomes reversed in midlatitudes, and in Iceland the ratio is 1:9. Even early stages in succession are quickly dominated by woody plants.

Some characteristic genera of lowland rain forest in Middle and South America are *Andira, Brosimum, Calophyllum, Castilla, Guettarda, Manilkara, Miconia, Pithecolobium, Pouteria, Terminalia,* and *Vochysia.*

The individuals representing each tree species are usually widely scattered. At most they occur in small family groups in a complex mixture of other species. In the Philippines a plot 0.13 ha in size included 100 species of woody plants exceeding 3 m in height. In Malaya one hectare contained over 200 species of trees 10 cm or more in diameter, despite the fact that many of them belonged to one family, the *Dipterocarpaceae.*[324]

This high dominance of woody plants over other vasculares is clearly a legacy from Cretaceous time, for nearly all Cretaceous angiosperm fossils appear to represent woody species. The primitiveness of the forest is further attested to by the fact that the percentage of diploids is highest here, and within groups that are represented both here and at high latitudes, those with the more primitive morphology occur in rain forest. During the long existence of this biota in a climatic belt of minimal stress, an astonishing array of intricate relationships between plants and animals, especially insects, has had time to evolve. At high latitudes the alternating stresses of cold or dryness seem to have placed a premium on adaptation to the physical environment, with drastic purges of contrasted nature frequently interrupting evolutionary trends throughout tracheophyte history.

The high species diversity among trees is probably related to the reproductive isolation conferred by nonsynchronous flowering and low dissemination efficiency, the former a consequence of the constantly favorable climate, and the latter a consequence of the need for disseminules well supplied with food to offset the severe competition seedlings must meet. Wide spacing of the individuals retards exchange of genes over the forest so autogamy is favored, and with negligible climatic restriction on nonadaptive variation, mutations have good chances of survival, which favors genetic drift. New taxa form in proportion to mutation rates (which are believed to be no more than average in the tropics) + receptivity of environment + time. The last two factors have been maximal in rain forests.[137] Possibly during successive periods when interglacial heat levels may have brought about fragmentation of rain forest, evolutionary divergence was also enhanced, but the fact that diversity is most pronounced at the genus level argues against this view.

The high level of taxonomic diversity in this vegetation provides an excellent illustration of the fallacy of considering competitive exclusion to the point of one species per niche a universal ecologic principle. Rather than there being 100 tree niches (time and/or space) in the 0.13 ha plot referred to previously, there are probably only a few niches, with chance determining which species gets established at a particular locus in each.[340]

Tropical rain forest is neither taller (rarely to 55 m), nor the degree of shade it casts more intense than temperate forests, but the typical tree differs in several respects. Commonly stature differs widely among the woody species, this resulting in a deep canopy with a very uneven surface, and sometimes distinct strata can be seen. The most nearly closed layer is usually composed of trees about 30–40 m tall. A few trees, the emergents, may have all of their canopies above this level, but seldom reaching as much as 55 m. Below the main canopy are shorter trees. Where strata are evident they are usually fewer than four, and commonly a gradient in the height of younger individuals and shorter species completely obscures stratification. Among the tallest trees the canopies tend to be broader than high, with the proportions reversed in the lowest species, and a gradient between.

Tree boles are usually not more than 1 m in diameter at breast height, often fluted, straight, and free of branches for most of their length. Branching orders are less complex (5 orders at most) than in temperate angiosperm forests. Upward thickening of lateral roots often produces wide-flaring buttresses that may extend as much as 9 m up the trunk and twice as far outward (Fig. 172). These are frequent, but by no means ubiquitous. They are features of upper and middle canopy species which lack tap roots, and appear to enhance stability.[189]

A scattering of palms which seldom reach the main canopy adds a striking bit of variety to community structure. Tree ferns and tree bamboos are common. Among understory trees, especially palms, adventitious prop roots are common. Both buttressing and prop roots tend to be more prevalent as soil aeration decreases, and they may be of real benefit to tall trees with such slender boles. Aerial roots descending from branches occur here as in seasonal forest.

Fig. 172.   Interior of lowland rain forest in Costa Rica.

The bark of rain forest trees is highly variable in character, but unlike temperate zone angiosperms, smooth-barked species are common. In many species the wood is too dense to float. However, young trees of the fast-growing seral *Ochroma lagopus* are the source of balsa wood, the lightest wood known.

Hard, specialized bud scales as on extratropical woody plants are absent, but embryonic leaves may be covered with tomentum or with stipular sheaths that drop off as the leaves expand. Young leaves frequently lack the green color of mature foliage, being reddish, white, or some other color, and while in this immature state they usually hang limply. Even when mature the shade of green varies considerably among the species. Nearly all the trees are evergreen, but they renew their foliage at intervals of about 6–12 months, often dropping their old leaves as the new ones expand.

Growth is not continuous despite the high degree of climatic uniformity throughout the year. Rest may come at regular intervals, with the cycles varying from 3–32 months in length. The relatively few species that stand leafless for a short time, mostly members of the tallest stratum, vary so much in the timing of their nakedness that the forest as a whole retains its ever-green aspect.[233] Often two trees of the same species have differing periodicities.[199] Even different branches of the same tree may exhibit contrasted behavior. Phenology appears to be regulated by an internal "clock" in some species, with slight chilling by a rain storm triggering the activity of others, and a brief rainless period affecting still others.[199] In all these patterns of behavior there is very little correlation with seasons.

In the upper tree layers species with compound leaves are abundant, but in the shorter species the leaves are mainly simple. The leaf blades, or leaflets, are mostly in the mesophyll size class in Raunkiaer's system of classification. Most are entire margined, and broadly elliptic with acuminate tips. The significance of the striking prevalence of entire margins is obscure, but there is some evidence that acuminate leaves shed water most rapidly, and this might tend to discourage parasitic invasion, or epiphylls, or it might reduce the amount of nutrients and soluble foods that foliage loses in rain storms. The leaves vary from thin to coriaceous, with their sizes and thinness varying inversely with the height of the mature tree. These xeromorphic features would seem to be a response to transpiration stress, since they are associated quantitatively with microclimate.

Flowering, fruiting, and consequently much insect and vertebrate activity, are concentrated in the tree canopies, much to the chagrin of earthbound biologists interested in studying reproductive processes or in obtaining specimens critical for taxonomic identification. In contrast with forests of temperate and subarctic latitudes, pollination by animals (insects, bats, birds) is the rule here, but even so the flowers are usually small, usually greenish or whitish, and not aggregated into conspicuous clusters. In many trees flower buds remain dormant beneath thin bark for many years, eventually breaking through to produce flowers on old wood. This phenomenon, cauliflory, is common only among the understory trees and lianas, and is usually associated with bat pollination or dissemination. With the individuals of each species so widely scattered, wind pollination would have to be very wasteful to be effective.

Except during violent storms there is very little air movement, hence, there is little dependence on wind for dissemination either, except among the tallest trees and among seral species. In general the fruits of rain forest trees are large, commonly fleshy, often have brightly colored valves or seeds, and so are attractive to animals. The seeds tend to be large and contain abundant reserves that help seedlings get established in weak light, but viability is ephemeral.

It is in these forests that lianas attain their best development, but they are not uniformly conspicuous throughout the Region. They are most abundant along river banks and in disturbed areas, where they tend to smother seedlings and saplings and thus slow the rate of secondary succession. Many are woody heliophytes (commonly *Bignoniaceae* and *Malpighiaceae* in Middle America) that spread their canopies over those of the tall trees and have stems hanging free as "monkey ropes" or "monkey ladders." Like their supports they are commonly wind disseminated. Other linanas are sciophytes [e.g., *Vanilla* (Fig. 173), *Monstera,* and *Philodendron*] that grow adnate to the tree trunks and live entirely in the shade of the main tree canopy. Thorny-leaved lianas of the palm family (*Calamus* spp.), whose slender stems undulate horizontally through tree canopies, sometimes reaching 100 m in length, are rather extreme examples of a common tendency for woody lianas to bind tree canopies together. In consequence, trees may not fall if cut singly, or conversely, an old tree may fall and pull down smaller neighbors to which it was bound by lianas.

Fig. 173. *Vanilla planifolia*, a liana climbing by means of root tendrils adhering to tree bark, remains in shade below the rain forest canopy.

Vascular epiphytes anchored on tree or liana stems, including herbs, shrubs, and even trees up to 15 m tall, are abundant in the American and Asian rain forests. In a given area their species commonly exceed those of the terrestrial herbs and shrubs. Long-lived leaves accumulate epiphyllous coatings of algae, lichens, and bryophytes. Among the epiphytes there is every gradation from hygrophytic types, especially filmy ferns (*Hymenophyllaceae*) and peperomias on the lower tree boles, to xerophytic orchids, bromeliads (Americas only), aroids, cacti, and lichens on the topmost branches. The abundance, diversity, and degree of xeromorphy in the epiphytic flora is a close reflection of the relative humidity of an area or of a microclimate.[381]

Parasitic *Loranthaceae* are abundant, and tree stranglers are represented by *Ficus* and *Clusia* in the Americas, and by other genera in the Old World.

Under the shade of mature forest there are many species but few individuals of herbs and shrubs, so that usually a person can walk about with no more impediments than in mature temperate zone forests. Herbs must here be defined only on the basis of a lack of woodiness of the stem, for plants of this texture may grow several meters tall and live for years, rather than dying back to the ground annually. In contrast with the trees their leaves are usually membranaceous. Grasses, though a minor element, are distinctive in having short, broad leaf blades that often bear superficial resemblance to those of the *Commelinaceae*. Small palms are common as shrubs. Many of these undergrowth plants have proven to be good house plants, owing to their tolerance of weak light, poor soil, and constant warmth. Furthermore, variegated leaves and species with relatively large flowers are common among these plants.

In early stages of secondary succession large monocot herbs in such families as the *Cannaceae, Marantaceae, Musaceae,* and *Zingiberaceae* grow very closely spaced among the rapidly growing seral trees. To the botanist mature rain forest is hardly a "jungle," but the term is appropriate for such seral vegetation, and for the narrow strips of edaphic climax forest along river margins.

Representative animals of rain forests of the Americas include tapir, peccary, dwarf deer, monkey, sloth, squirrel, large rodents such as the capybara, cat (jaguar, puma, ocelot, margay), bat, parrot, macaw, curasow, pigeon, vulture, eagle, toucan, tanager, hummingbird, tinamou, and many snakes and lizards (iguana, crocodile, caiman). Owing to the prevailing wetness of lowland rain forest, crabs and frogs (and in Asia, leaches) are not restricted to aquatic or riparian habitats. Termites and ants abound, the latter often occupying special cavities in stems and leaves, even in the upper canopy. For the most part, the character of the faunas is quite similar in lowland forests, montane forests, and seasonal forest.

Classification

In the Tropical Mesophytic Forest Region edaphic, topographic, and fire climaxes are generally dominated by one or a few species, and their classification poses no special problem. However, the classification of climatic climaxes is difficult. First, there appear to be many species with closely similar environmental requirements that must be considered as ecologic equivalents. Second, community structure is quite sensitive to topography and soil, and since these vary in short space, homogeneous areas are not large. Thus, the key sociologic parameters of density and dominance for each species seem usually to require larger tracts for their approximation than topographic and edaphic variation provide. Compounding the problem is the ubiquity of past cultivation by primitive peoples, combined with the centuries required for a return to stability.[338]

The species-area curves that have been presented show no tendency to flatten even when as much area as 1.6 ha have been tallied, and for this reason some have concluded that there is no significant degree of discontinuity in rain forest on zonal soils. However, the curves are usually based entirely on trees 10 cm in diameter or larger, and this alone could prevent flattening.[323] Furthermore, there is no convincing evidence that the species-area curves that have been presented were obtained using an area homogeneous as to soil, topography, and past disturbance. Only recently has the ubiquity of disturbance been realized. In western Africa all rain forest appears to be a mosaic of seral communities of differing ages, and it is frustrating to attempt their classification without knowing the character of

undisturbed vegetation.[338] Much could be accomplished by emphasizing
population structure, but this has rarely been attempted.[355] Finally, herbaceous and shrubby vegetation has thus far been almost entirely ignored in
classification, and this element may prove as highly useful for indicating
ecologic discontinuities as it has proven in temperate latitudes.

A number of experienced investigators[9,25,355,398] have concluded that
there are indeed significant discontinuities in rain forest vegetation. According to P. W. Richards "The problems of tropical ecology appear somewhat
different from those of temperate regions, but are fundamentally similar."[339]
He had this to say about virgin rain forest in Guyana which he had studied in
detail: "There are a number of clearly differentiated climax plant associations... which are readily recognized and defined. They could be
mapped and their characteristics can be precisely stated."[338] According to A.
Gomez-Pompa "one of the striking things about these... high evergreen
selvas... is that they tend to form floristic units locally that can be distinguished rather easily by a person well acquainted with the area."[157] Owing
to the greater species diversity here in comparison with temperate zone
forests, it is undoubtedly necessary to recognize more polymorphic associations just as species taxonomy must recognize very polymorphic units in
certain genera.

Nonzonal Vegetation[402]

Distinctive forest types are often associated with high water tables combined with oligotrophic water, and in these places woody peat accumulates,
with deposits up to 30 m thick reported from Malaya.[424,439] The peat surface
may be convex with rain the sole source of its moisture. In the central part of
these bogs the trees are shorter and the species fewer. Buttressing and
pneumatophores are common among the dicots of the tallest stratum. Palms
and epiphytes are abundant, but bamboo, tree ferns, lianas, and undergrowth are sparse. The water draining from such ecosystems is coffee-
colored from humus colloids (as in the Rio Negro of Brazil) and is highly
acidic. Such bog forests occur over wide area in South America and Malaya.

Along the landward margin of mangrove, surface drainage may be sluggish
but the waters eutrophic, therefore peat does not accumulate. Swamp forests
are dominated by a few specialized trees which may be either palms or dicots
with conspicuous buttresses and pneumatophores. Like rain forest, these
woody plants are evergreen, and may form distinctive strata. Again the low
potential evapotranspiration of such wetland ecosystems is reflected in an
abundance of epiphytes and lianas.

Large rivers commonly overflow their banks for several months each year
when rainfall is heaviest, thus creating seasonal freshwater swamps. In the
Amazon basin where river levels rise about 20 m during these months,
special riparian forests are many kilometers wide.

Marshes occur sparingly in the Rain Forest Region, and where the water is deep enough they may have floating mats of vegetation.

Mangroves along the coasts in ever-wet climates do not differ significantly from those bordering seasonal forest.

In Surinam, savannas have been described that occur on silica sand that is low in iron and other plant nutrients, and dries quickly during the brief rainless periods.[191] Elsewhere special communities on silica or granitic material are associated with podzol profiles with a cemented ortstein up to 2 m thick and yielding acid brown drainage waters.

In places where man has abandoned cropland owing to weedy grasses, and subsequently burned the grasses at intervals, derived savanna has developed. In Guatemala it has been alleged that repeated burning, not necessarily preceded by cropping, has resulted in *Curatella–Byrsonima* savanna in rain forest habitats.[264] The soils of such savannas stand in need of careful investigation, for very different interpretations might be appropriate.

Land Use[213]

Unmanaged rain forest is the source of several commercial products, including Brazil nuts (*Bertholettia excelsa*), chewing gum (*Manilkara zapota*), cola nut (*Cola* spp.), mahogany (*Swietenia macrophylla*) and fine cabinet woods in other genera, quinine (*Cinchona acuminata*), rattan (lianas in *Calamus*), rubber (*Hevea brasiliensis*), vegetable ivory (*Phytelephas macrocarpus*), fustic (*Chlorophora tinctora*) and other dye stuffs, gutta percha (*Palaquim*), kapok (*Ceiba pentandra*), and carnauba wax (*Copernicia*).

Long before Caucasians explored rain forest, primitive man had developed independently a successful type of agriculture in nearly all parts of the lowland rain forest, and in seasonal forest as well. This involved clear-cutting forest on a small tract of zonal soil, burning all the plant debris to make available its mineral content, and then planting a crop immediately. The fertility thus released in the ash exceeds the ability of the small population of crop plants to take it up promptly, consequently much of it is lost by leaching in addition to that lost in runoff and taken away in the crop. Furthermore, the continual nutrient replenishment from litter cast by the heavy forest cover is eliminated, and at the same time warmer soil temperatures increase the rate of decay of the slender humus supply. Weeds become increasingly more troublesome after forest is removed, as do insect and fungus pests which have no cold or dry season to check their continual increase.[422] As a garden plot deteriorates it is abandoned in favor of a new clearing made elsewhere. Although this cropping system provides an abundance of starchy crops such as cassava (*Manihot utilissima*), sweet potato (*Ipomoea batatus*), taro (*Colocasia esculenta*), yam (*Dioscorea alata*), banana (*Musa sapientum*), and plantain (*Musa paradisica*), the mineral and protein content of such food is low, and it is necessary to depend on gathering fruit and nuts, as well as hunting and fishing to supplement these garden products.

If the abandoned fields are burned frequently the grasses that have been among the increasingly troublesome weeds, especially *Imperata, Panicum* or *Pennisetum,* can become dominant in fire climax savanna. But if abandoned fields are not burned they are quickly covered by root sprouts from woody plants that were not killed when the preceding forest was cut and burned, and have survived weeding activity. Joining these survivors at once are seral invaders with highly mobile disseminules. Early seral vegetation in Middle America includes large-leaved herbs (e.g., *Calathea* and *Heliconia,* the latter to 5 m tall), together with rapidly growing soft-wooded trees such as *Cecropia peltata, Ochroma lagopus,* and *Trema micrantha,* some of which may mature and be eliminated in 15–20 years. Early seral communities, with many stems per square meter, often bound together with lianas, form an almost impenetrable thicket—the "jungle" of plant geographers.

Only 10–30 years are required for seral vegetation to accumulate enough nutrients, and for pest populations to subside, so that another round of cutting, burning, and cropping is practical. Despite the rapidity with which this seral vegetation develops, centuries are needed for rain forest to regain stability, since the trees of intermediate stages are long-lived, and climax species are slow to reenter and adjust to one another. In the 500–600 years since the temple of Ankor Vat was built (with sun-dried laterite used for bricks!) the surrounding rain forest has not yet achieved the normal character of undisturbed areas.[430]

Thus, we see that shifting agriculture keeps large areas of forest in varying stages of secondary succession, but on gentle slopes (Fig. 174) it is nevertheless ecologically sound in that the brief cropping periods do not progressively degrade the soil as where the land is cultivated year after year. The major limitation is that zonal soils can support only about 10 persons/ha in this life style, and any tendency to shorten the period of fallow or lengthen the period of cultivation in order to increase food output initiates feedback leading to disaster. It is suspected that the demise of Maya civilization in the rain forests of Middle America may have resulted from agricultural failure associated with the development of a high population density, leading to a shortening of the fallow period.

In view of the risk involved in bringing lateritic soils under cultivation, the safest way of increasing the level of their productivity above that of shifting cultivation appears to lie in simulating natural ecosystems. Natural forest can be replaced with artificial forests of economically valuable woody plants,[121] preferably rich mixtures involving more than one layer, so the land would have to be cleared for replacement only at long intervals, and then only briefly. Such plantations might include various combinations of the following: abaca or manila fiber (*Musa textilis*), avocado (*Persea americana*), banana (*Musa sapientum*), Brazil nut (*Bertholettia excelsa*), breadfruit (*Artocarpus*

Fig. 174.   Scarcely any slope is too steep to deter shifting agriculture where people are hungry. Rain forest Region near Valles, Mexico.

communis), cacao (*Theobroma cacao*), cashew nut (*Anacardium occidentale*), citrus fruits (*Citrus*), cola (*Cola*), custard apple (*Anona*), grenadilla (*Passiflora edulis*), guava (*Psidium guajava*), jack fruit (*Artocarpus integer*), jute fiber (*Corchorus*), macadamia nut (*Macadamia ternifolia*), mango (*Mangifera indica*), oil palms (*Cocos nucifera, Elaeis guineensis*), papaya (*Carica papaya*), pepper (*Piper nigrum*), plantain (*Musa paradisiaca*), rubber (*Hevea brasiliensis*), rotenone (*Derris*), vanilla (*Vanilla planifolia*), wax palm (*Copernicia cerifera*). The hazards of growing such plants in monocultures is well illustrated by the disastrous collapse of the banana industry on the east coast of Central America when a fungus disease was accidentally introduced.

In certain oriental countries terraced hillsides are used as permanent fields for cultivating paddy rice or taro, but irrigation, nutrient cycling (augmented by nitrogen-fixing bluegreen algae stimulated by human wastes), and a high manual labor input are required.

Modern agricultural technique has been successful in the wet tropics where young volcanic soils have not developed into mature laterite. Thus, pineapple and sugar cane are regularly cultivated on such soils in Hawaii. But nutrients are lost so rapidly by leaching that fertilizer must be applied frequently in small amounts, which greatly increases the cost of production. Also it is difficult to prevent rapid erosion where the soil is cultivated annually without terracing, and the uninterrupted multiplication of insect and fungus pests complicates control measures.[422]

Floodplains also provide soils that can be cultivated annually to yield crops that compare in quantity and quality with those of temperate latitudes. Here fertility is annually renewed by silt deposited as rainy season floods subside, and then crops are planted immediately. Large deltas and wide floodplains are available for such use, and even primitive methods have allowed annual cropping with high levels of productivity that have supported high population densities. Temperate zone engineers, obsessed with the notion that their own life styles are ideal, have suggested that flooding of the Mekong River should be controlled by dams and levees. If this were done it would disrupt the life styles of people well adapted to their environment, and make the importation of fertilizer necessary, with no certainty of increasing productivity.

Lower Montane Rain Forest[27,89]

Since low temperatures reduce the significance of a dry season, rain forest may occur at high altitudes even where there are several months of low rainfall (Fig. 14B). There, heat deficiency may become of sufficient importance that phenology tends to be synchronized by the small annual variation in monthly temperatures.

Although the climates remain in the ever-wet category, and most of the characteristic animals of the lowland extend upward, the character of rain forest vegetation changes progressively with altitude above a basal plain. Accompanying the progressive drop in temperature: (1) floristic diversity decreases; (2) forest stature declines and stratification becomes simpler, with the main canopy becoming smoother; (3) tree buttressing and cauliflory decline sharply; (4) lianas usually diminish and become limited to chiefly the sciophytic class; (5) epiphytes, especially ferns and bryophytes, increase greatly; (6) the shrub layer is better developed; (7) palms decrease in representation, but shrubby bamboos commonly form thickets; (8) tree leaves become smaller, less acuminate, more frequently nonentire, thicker and more firm-textured, with this xeromorphy most likely a reflection of poor mineral nutrition,[165] possibly combined with a transpiration stream inadequate to conduct nutrients upward; (9) the rate of litter decay decreases so that surface accumulations of litter and duff become progressively thicker, with more and more of the nutrient capital immobilized in detritus; (10) soil pH declines and humus content increases; (11) soil erosion is more active than on the basal plains, hence parent materials play a more important role in vegetation differentiation.

Although the foregoing characters tend to change gradually upward, at least for a distance, certain features of climate and plant life serve to distinguish three forest belts above lowland rain forest, wherever mountains rise high enough. Many workers have recognized a vegetational discontinuity marking the upper limit of lowland rain forest at elevations ranging from 250–1500 m in different geographic areas, the position of this ecotone being higher the larger the mountain mass or the drier the lowlands. Lower montane rain forest lies above this ecotone.

The soils at this elevation are not laterized, so modern agriculture can be practiced without serious hazard of soil deterioration other than erosion, the chief crops being, e.g., coffee (Fig. 175), tea, grains, vegetables (beans, peas, white potatoes, beets, carrots, cabbage, squash), pasturage for dairy cattle, and sisal. The shift from warm climate suitable for cacao to cool climate suitable for coffee often provides an agronomic base for locating the division between lowland and montane rainforest. Aborigines practiced shifting agriculture here to only a limited extent, but in the developing countries this practice has been expanding rapidly owing to population pressures from below. Caucasians find the climate in this lower montane rain forest belt most delightful, and for this reason it has been selected for the location of

Fig. 175. Coffee bushes grown in the shade of leguminous trees in lower montane rain forest in Nicaragua. Usually, but not here, the trees are kept lower by heavy pruning each year, with the debris left to decay on the ground.

many tropical cities, so long as there is no impelling reason for locating them on the basal plains.

Midmontane Rain Forest (Fog Forest)[65,277,303,304,362]

With increasing altitude and a continuation of the vegetational trends listed previously, a level is reached where the mountain sides intercept the normal cloud stratum, so that the vegetation is swathed in fog much of the time. Although one of the highest mean annual rainfalls in the world has been reported from a midmontane location in the tropics (11,430 mm on Mt. Waialele, Kauai, Hawaii), rainfall on other tropical mountains is not necessarily highest in this fog belt. Usually the rainfall is not distributed evenly through the year, but there is a constant or near-constant moisture surplus resulting from tree foliage combing droplets out of the drifting fog. In Ecuador where rainfall is quite seasonal, even during the "dry" season it was noted as being unusual for as much as 2 weeks to pass without fog.[166] In the Philippines the lowest relative humidity recorded by another observer was 91%. These foggy environments may occur on mountain slopes above arid basal plains,[406] but the elevation of the belt is higher. Where mountains rise into but not above this fog belt, deforestation allows much water to pass beyond the area where trees would otherwise intercept it, thus reducing water supplies that may be needed on the basal plain below. Reforestation, or at least the erection of screens as mist traps, could provide valuable increments to stream flow in such places.

Abundant foliage drip combined with very low evapotranspiration keeps the soils so wet that frequently a gley layer develops. Elsewhere podzols with thick litter and duff are common, and local patches of *Sphagnum* may occur on the surface. Termite and earthworms cannot thrive here as below fog forest.

In the Americas fog forest extends from Mexico to Peru. In South America the belt is often referred to as the "ceja de la montaña." "Cloud forest," "mist forest," and "mossy forest" are other names which have been applied. In dry areas the fog belt is lacking and this segment of montane forest is absent.

Characteristic genera north of the equator in the Americas are *Clethra*, *Clusia*, *Cyrilla*, *Drimys*, *Engelhardtia*, *Eugenia*, *Oreopanax*, *Persea*, *Podocarpus* (a gymnosperm), *Psychotria*, *Quercus*, *Rapanea*, *Ternstroemia*, and *Weinmannia*. Progressing northward through Central America to Mexico these genera drop out one by one with *Quercus*, *Pinus*, and *Cupressus lusitanica* gaining representation. The northern extremity of midmontane rain forest contains the disjunct outposts of temperate mesophytic forest that were discussed previously. Similar floristic trends characterize the mountains of southeastern Asia.

The land discontinuities between Mexico and Colombia that hampered intercontinental migrations from Eocene to late Pliocene time not only kept

Pinus and *Cupressus* from extending south of Nicaragua, but kept montane genera such as *Heteromeles* and *Oreopanax* from migrating north of Costa Rica. However, *Quercus* was able to surmount the water barrier and extend from its area of major concentration in Mexico to northern South America, while *Podocarpus* crossed in the opposite direction and has reached as far north as Mexico. Many of the peaks that rise barely into the cloud level are so widely scattered that endemism is quite high in the flora.

In some areas the tallest trees have buttressed bases. Beneath them a thin second stratum of tree ferns (*Alsophila, Cyathea*) is sometimes conspicuous (Fig. 176). Among the abundant shrubs the large-leaved *Gunnera* is especially noteworthy. Forest floor herbs include *Begonia, Calceolaria, Dahlia,* and *Fuchsia*. Although lianas are generally lacking, thin-stemmed species of bamboo, chiefly *Chusquea* in the Americas, frequently occur as scramblers, or as erect plants forming clonal thickets. Palms are often lacking, but in Puerto Rico the palm *Euterpe globosa* dominates a large area in this belt.

Nowhere else in the world are vascular epiphytes (orchids, bromeliads, aroids, and especially ferns) as well developed as here (Figs. 99, 177, and

Fig. 176.   Tree fern in midmontane (fog) forest. Henri Pittier National Park, Venezuela.

Fig. 177.   Epiphytes on the lower trunk on a tree in midmontane rain forest, Tamazunchale, Mexico.

178), and this character serves best to distinguish midmontane rain forest. A single tree in Sumatra proved to have over 50 species of ferns alone growing upon it. Also, bryophytes and filmy ferns are often extremely abundant, if not dominant. Such epiphyte loads may become so heavy when water soaked as to break off tree limbs.

An especially interesting variant of fog forest occurs on windswept ridges and peaks that extend into but not above the cloud stratum. Here the stature of the vegetation is low (2 m or less in places) in proportion to the amount of wind, and the canopies are of conspicuously uniform height. Although the dense canopy surface appears wind shorn, there would seem to be no possibility of the involvement of a water deficit with fog so plentiful. In Puerto Rico 100% of the night hours and 60% of the daylight hours proved foggy on such a ridge. However, there still are occasional brief periods of sun, and if these are combined with wind, it might well account for the dwarfing.[303]

This low one-storied community is rendered nearly impentrable by the high tree density, branching near the ground, and frequency of adventitious

Fig. 178.   Tree branches here in midmontane rain forest have accumulated a heavy load of epiphytes, among which the *Bromeliaceae* is conspicuous.

roots emerging laterally from the trunks or dropping from the branches. Owing to the gnarled trunks, crooked branches and low stature, this variant of fog forest is sometimes referred to as "elfin wood."

Litter and duff may be as much as 35 cm thick, with herbs and shrubs essentially lacking. Where wind is not excessive, bryophytes and *Selaginella* may be copiously represented both on the ground and on tree stems. Vascular epiphytes too may abound.

Land use in the fog belt is limited, and the environment attracts few permanent residents. However, the pressure of excessive populations has begun to show even here, with forest clearings made for pasture.

Upper Montane Rain Forest[404]

Where mountains rise above the prevailing cloud stratum, the climate is usually still moist but it is cold enough for light frost on occasion, and the skies are clear. Undisturbed vegetation consists of low forest just above the cloud level, grading into dwarfed trees at the upper timberline, which in turn give way to a belt of shrubbery that is usually considered the lowest member of the alpine series.

From Costa Rica southward members of the *Ericaceae* are especially conspicuous among the dominants—*Clusia, Gaultheria, Escallonia, Arctostaphylos, Heteromeles, Pernettia, Psychotria*. The *Melastomaceae* is also well represented, and here, in contrast with fog forest, genera common to temperate latitudes are frequently encountered.

Not only are the trees low, but their stems are gnarled and frequently intertwined, and the canopies tend to be dome-shaped or flat. On ridges or peaks that extend no higher than this belt the vegetation may be dwarfed until scarcely over a meter in height. The evergreen leaves are frequently rusty pubescent, and provide a very dense shade.

Trees reach their highest elevation in the bottoms of ravines, in marked contrast with those of mountains in temperate latitudes where ravine bottoms tend to be frosty and treeless from cold air drainage. This topographic restriction does not seem to be related to wind, or at least to winds from a constant direction, for the trees are symmetrical, and show no response to other kinds of microsites that provide wind shelter.

Organic matter of the podzolized profiles is thinner than in the fog forest, and burning has been widespread since the environment is not so wet. Where the fog forest belt is lacking, cultivation may extend to upper timberline.

In central and northern Mexico the upper montane rain forest is replaced geographically by the subalpine *Pinus hartwegii* belt.

TROPICAL ALPINE REGION

Tropical mountains high enough to exceed tree limits occur in Middle and South America, Africa, Australia, Borneo, New Guinea, and Hawaii. From Nicaragua north the alpine floras of the Americas have very few tropical genera and species, and for that reason they have been treated with extratropical alpine tundra.

The Andes uplifted sufficiently to support alpine vegetation at the end of the Pliocene Epoch. Owing to the isolation of South America at that time, the floras of each mountain system had to be derived from the taxonomic stocks of the surrounding lowlands, or from immigrants coming down the cordillera from the north temperate zone. Endemism is accordingly high. During at least the last glacial episode of the Pleistocene Epoch, timberlines in South America were lowered approximately 1100 m. Fossil stone nets and stripes are common, as are glacier-scoured cirques.

Where the mountains are high enough, the tropical alpine zone occupies about 1000 m of elevation between upper timberline and permanent snowfields. Mean annual temperatures range downward from about 10°C, and with negligible seasonal differences, the growing season, though chilly, is still 12 months long (Fig. 14C). Frost or near-frost conditions occur every night, with daytime temperatures rising sometimes to 20°C in the lower part of the belt. This temperature regime has been aptly described as "winter every night and summer every day!" Such a wide daily range, coupled with negligible difference between summer and winter, contrasts sharply with Arctic and Antarctic tundras where there is little diurnal variation but strong difference between summer and winter. Another striking contrast is that in the tropical alpine belt the limited snow showers that fall on the vegetated areas thaw promptly, so that special snowpatch communities are not found here.

Upper timberline is lower in equatorial latitudes than in the subtropical high pressure belts to both the north and south, presumably on account of more cloudiness and consequently a reduced heat supply.[100]

As in other alpine areas, violent chilly winds are common.

In northern South America,[415] high rainfall and dew keep the alpine soils constantly moist if not wet. Some soil profiles are deeply humified, with the horizons feebly differentiated. A Páramo Humus Soil described in Colombia has a sandy, acid, black to dark brown upper horizon, with gley below.[215] Elsewhere podsols or peat accumulations are found.

The lower border of the alpine vegetation often consists of evergreen scrub (Fig. 179). These shrubs have mostly small, pubescent leaves, and include species in *Berberis, Gaultheria, Gaylussacia, Hesperomeles, Hypericum, Ilex, Pernettia, Rubus, Senecio,* and *Vaccinium.*

Above the scrub belt the dominants are caespitose grasses with coarse, stiff leaves—a marked contrast with high-latitude tundras where fine-textured swards with abundant *Cyperaceae* are the rule. In the Americas these coarse grasses include *Aristida, Calamagrostis, Danthonia, Deschampsia, Festuca, Koeleria, Muhlenbergia, Sporobolus, Stipa,* and *Trisetum.* Since the continued growth of the grass is accompanied by the continuous death of old leaves that remain standing for a long time, the herbage appears drab and brownish. Forbs in this grass-dominated layer also include genera common in the north temperate zone, e.g., *Castilleja, Draba, Erigeron, Gentiana, Geranium, Lupinus, Ranunculus,* and *Stellaria.* With so many genera shared with the north temperate zone yet lacking on tropical lowlands, it appears that there has been a sizeable migration southward down the cordillera since the Andes uplifted.

The ground between the grasses and forbs is occupied by an abundance of dwarf shrubs, cushion and rosette plants, and mosses, these adding two more strata to the vegetation. Then, in addition to the grass layer, there are often scattered evergreen shrubs of a highly distinctive and bizarre growth form, these mostly having a rosette of large, white- or brown-wooly leaves that are either sessile or are borne on the summit of a thick and mostly

Fig. 179.   Alpine scrub 1–2 m tall in Páramo Negro, Venezuelan Andes.

Fig. 180.   *Espeletia schultzii* in páramo at Lago de Mucubaji, Venezuelan Andes.

unbranched trunk (Fig. 180). Plants of this growth form are furnished by *Espeletia* (*Compositae*, having about 30 species, and almost entirely in the alpine vegetation), with *Blechnum, Culcitium, Lupinus,* and *Puja* locally common. In some species of *Espeletia,* trunks may lift the rosette to 4 m above the ground. In Hawaii the endemic *Argyroxiphium sanwicensis* represents the type, in Indonesia *Anaphalis* provides a semidendritic equivalent, and in Africa *Senecio* and *Lobelia* have physiognomically similar species in the alpine vegetation.[185] It is remarkable how this growth form has arisen in so many isolated places in the tropics around the world!

In the same middle alpine belt with the grasses and forbs and scattered tall shrubs, there are wet habitats with compact cushion plants (in *Azolla, Distichia, Draba, Geranium, Plantago,* etc.) much larger than any cushion plants encountered in extratropical latitudes. Some of the individuals cover several square meters of area.

In the third distinctive belt of the Tropical Alpine Region, the coarse forbs drop out leaving what were the subordinate layers below as the most conspicuous elements of the phytocoenoses. The vegetation is more open, and mosses and lichens play a more important role here. The soil is usually coarse and stony, and nocturnal snow up to 2 cm deep is frequent, although it almost invariably melts away the following day. At the upper limits of this belt the plant cover thins out rapidly as strong freeze–thaw cycles involving the formation of needle frost each night prevent seedling establishment, so the soil remains mostly bare between the stones.

Bogs and fens are common features of tropical alpine landscapes, as in tundras of higher latitudes. Animal life includes white-tailed deer, hare, bear, wolf, and condor.

In northern South America the land above tree growth is wet from the combined effects of rain, fog, and cloud the year around, in addition to being chilly, and so is hardly fit for human occupancy or any land use other than grazing. Frequent burning with the intent of eliminating old dead grass leaves to improve grazing has altered the character of most of the lower and middle alpine belts, and by destroying the margin of the upper montane forest has allowed scrub to extend downward. The alpine vegetation in this part of the cordillera is commonly called páramo. Native animals include the tapir, dwarf deer, spectacled bear, condor, thrush, wren, and hummingbird.

In that segment of the Andes extending south from Ecuador the alpine climates include an increasingly longer rainless summer. Between approximately 23–28° south latitude, where desert of the subtropical High Pressure belt occurs on the basal plains, the Andes, rising even as high as 4500 m, cannot stimulate enough precipitation for forest to grow at any altitude. There, in the cold windy climate of the alpine belt, is a broad tract of grassland locally called puna. Dominant grasses are provided by *Calamagrostis, Deyeuxia, Festuca, Poa* and *Stipa,* associated with endemic forbs in *Arenaria, Astragalus, Draba* and *Gentiana.* Locally large caulescent plants of

Puja (*Bromeliaceae*) lend variety to the drab landscape. On stony soils there is a scattering of large, compact cushion plants up to a meter tall. As in the northern Andes, disjunct groves of *Polylepis sericeus* trees occur on talus slopes and coarse alluvium well above the general thermal limits of forest.

The native fauna includes guanaco, vicuña, deer, cony-like rodents called chinchilla and viscacha, Andean fox, puma, condor, vulture, tinamou, hummingbird, flamingo, and a number of small birds which dig burrows in the soil.

Man has long lived in permanent villages in this part of the Andes. At the time of the Caucasian invasion, the Incan people living above tree limits were growing white potato (*Solanum tuberosum*) and to some extent tuberous-rooted species of *Oxalis* and *Tropaeolum* as their basic food. Tubers were exposed so as to freeze at night, the free water squeezed out after warming under the sun, the tissues then were dried for storage, a practice that still continues. As beasts of burden and sources of wool these people had domesticated the alpaca and llama from the native vicuña and guanaco. European sheep were introduced later.

VEGETATION IN RELATION TO CLIMATIC CLASSIFICATION

The Perspective

The concept of macroclimate as the master factor governing the gross pattern of the earth's terrestrial vegetation was emphasized by Alexander von Humboldt in 1805. Perhaps it was with the idea of testing this hypothesis that he keenly felt the need for the systematic collection of weather data, and his efforts to promote governmental interest in the undertaking finally bore fruit. But it was not until the midnineteenth century that there were enough data available to attempt quantitative studies of climate in relation to vegetation. Toward the end of the nineteenth century a related idea was formulated—that soils likewise reflect climate, so the patterns of vegetation, soil, and climate should all be interrelated.

Climate stands in peculiar relation to vegetation and soil patterns in that its gradients are gradual, so that any classification of climatic types based on physical data alone would be mainly arbitrary. Thus, isohyets are commonly made to show areas of 0–50 mm of rainfall, 50–100 mm, etc. Vegetation and soil categories, on the other hand, are less of an abstraction. They present reasonably distinctive types owing to (1) rather definite limits to the environmental tolerance of dominant plants, and (2) to vegetation influences on soil development. Climatic climaxes permit the delimitation of fairly homogeneous climatic areas, each tending to have distinctive soil characteristics. Descriptive climatology thus has traditionally looked to vegetation patterns for the location of points along climatic gradients that distinguish areas having the most biogeographic significance.

Synecology is not only of service to quantitative climatology in the above manner, but climatology is, in turn, useful in synecology, for the relations that can be established empirically then become clues to factors which have had the greatest influence in determining the boundaries of present vegetation types, as well as influencing past evolution and migrations. From the biologist's standpoint the value of a climatic classification is judged by its closeness of fit to the distribution of climatic climaxes.

Some Historic Landmarks[176,251,392]

The Greek System of Climatic Classification

During the sixth century BC the Greek philosopher Parmenides of Elea advanced a concept that was to prevail for over 2000 years, namely, that the earth possessed five climatic zones: one Torrid Zone flanked by two Temperate and in turn by two Frigid Zones. Temperate was distinguished from Torrid by the Tropics of Cancer and of Capricorn, which represent the farthest points from the equator where the sun is directly overhead at noon on the longest day of the year. The Frigid Zones were distinguished from the Temperate by the Arctic and Antarctic Circles, where the center of the sun is barely visible on the horizon at noon on the shortest day. Although they surmised that the earth was round, the Greeks thought that the sun came so close to the surface at the equator that a barrier of intolerable heat would prevent man from crossing the Torrid Zone, whereas the Frigid Zone was too cold to be inhabitable.

Astronomic data provide the entire basis for this indirect climatic classification, and although it reflects the theoretical temperature gradients from the equator to the poles, the result is extremely crude in the face of great irregularities in the distribution of heat. The moisture component of climate was completely ignored. The Greeks had no experience with areas of high rainfall. Neither were they sufficiently traveled as to sense the discrepancies between temperature and latitude. Obviously the concept of a close relationship between vegetation and climate was yet unborn.

Humboldt: First Use of Weather Data

The invention of the thermometer in 1714 made it possible to start gathering quantitative data and eventually to compare areas on this basis. However it remained for the idea to spread and to be put into practice for sufficient time that reasonably reliable averages of temperature conditions could be derived. Until such data representing a geographic spread of locations became available, the classic concept of astronomically based thermal belts had to prevail.

Alexander von Humboldt was among those in the early part of the nineteenth century who, following the Greek perspective, regarded temperature the most important aspect of climate. In 1817 he assembled data and located isotherms of mean annual temperature on a map. This showed the great advantage a map has over a tabular presentation of data. Since Humboldt's field experience had been mostly in the tropics where there is little seasonal variation in temperature, it is not surprising that he considered annual means satisfactory expressions of temperature relations.

Meyen: The Importance of Seasonality

In his treatise on plant geography published in 1836, F. J. F. Meyen presented graphs showing variations in mean monthly temperatures through the course of the year in extratropical latitudes. These were the basis of his emphasis on the fact that annual means have very low value for expressing extratropical temperature conditions in relation to plant geography. Meyen recognized rainfall as a climatic element important in plant geography, but as yet this aspect of climate was not quantified.

Linsser: Quantification of Precipitation Effectivity

In 1869 Carl Linsser proposed a method of evaluating the effectivity of a given amount of precipitation by calculating the ratio of monthly precipitation to monthly temperature.[1] On this basis he subdivided the earth into six zones. This concept of discounting precipitation in proportion to temperature was so far ahead of its time that its significance was overlooked for several decades. In fact, his contribution seems never to have been acknowledged by subsequent workers in this field.

De Candolle: Recognition of Arid Climates

In 1874 the plant physiologist Alphonse L. P. P. De Candolle, son of the famous taxonomist Augustin P. De Candolle, presented a classification of five major types of land plants according to their climatic requirements: (Although this classification was autecologically oriented, by implication it recognized major latitudinal belts of plant life which included a xerophytic member in the latitude of the Subtropical High Pressure belt.) (A) *Megatherms*, tropical plants requiring a mean annual temperature between 20–30°C and abundant moisture; (B) *Xerophiles*, plants mainly of the Subtropical High Pressure belt that are tolerant of a regular season of severe dryness; (C) *Mesotherms*, plants requiring a mean annual temperature of 15–20°C and abundant moisture; (D) *Microtherms*, plants requiring a mean annual temperature between 0–14°C and abundant moisture; and (E) *Hekistotherms*, plants growing where the mean annual temperature in less than 0°C.

De Candolle either did not know of Linsser's contribution or did not appreciate its significance. Neither did he grasp the importance of seasonal differences in temperature as Meyen had pointed out. But he agreed with Meyen that more emphasis should be given the moisture aspect of phytoclimate and suggested that the number of rainy days per month would serve as a good criterion of the degree of adequacy of moisture. Such data were, however, still not sufficiently abundant to be useful in plant geography.

Köppen: Quantification of Seasonal Moisture and Temperature, and Polynomial Symbolism

In 1900 Vladimir Köppen proposed a climatic classification which attempted to improve upon De Candolle's expression of plant types, and create a universal quantitative classification of climates that would fit vegetation types. He started with maps which plant geographers had made showing the world's pattern of major vegetation units, and sought ways of equating the ecotones with points along moisture and temperature gradients. Annual means were mostly avoided, for as he said, in them "the most diverse situations are buried together," and, in consequence, he stressed the changing seasonal relationships between temperature and precipitation, although he did not combine the values arithmetically as Linsser had done. A major innovation was introducing polynomial symbolism so that different aspects of climate could be expressed independently. The climate of Wapato, Washington, for example, is BSks.

Köppen expended much effort in selecting limiting values. Some of those which he considered critical were as follows. A mean monthly temperature of −3°C was taken as the highest temperature at which a snow cover can persist and give protection to low-growing plants. A mean of 10°C for the warmest month of the year was considered as approximating the cold limits of tree growth, and a mean temperature 18°C for the coldest month marked the limit of cold tolerance of tropical plants. Temperature criteria such as these were used to redefine the A, B, C, D, and E climates of De Candolle, and the second member of the polynomial indicated the seasonal distribution of rainfall. The third and subsequent letters could be added as appropriate to indicate the level of summer heat, continentality of climate, frequency of fog, etc. When mapped, such climatic types were shown to recur in different parts of the world and exhibit a symmetrical pattern that did not conform to strictly latitudinal belts, but did conform well with the best map of the world's vegetation, which had been prepared by the plant geographer A. H. R. Grisebach in 1866.

Köppen achieved great fame as a result of this classification, however its faults are not difficult to see. All values he considered critical were strictly empirical rather than resting on the results of ecologic research, and when applied to a limited area show many discrepancies between the vegetation pattern and the classification. Nevertheless, the classification provided a rough index of climatic equivalence, and the popularity which it gained after his 1918 modification appeared has persisted to the present time.

Thornthwaite: A "Rational" System

In 1931 C. W. Thornthwaite proposed a climatic classification which embodied the best of Köppen's ideas (polynomial symbolism, and emphasis on the seasonal variations in water and heat) and added some improvements. He strove to emphasize the efficiency of moisture and temperature in promoting plant growth. Here for the first time water balance was given precedence over temperature. The polynomials he used consisted of five letters, these indicating five categories of precipitation effectivity, six categories of temperature sums, four patterns of annual distribution of precipitation, and five categories for the percentage of heat received in the three summer months.

A major deficiency of this system was his summation of monthly precipitation-effectivity calculations to obtain an annual value—a procedure that can result in the same sum for wide differences in the series of values added together. Secondly, the range of values representing each climatic characteristic was divided into rather arbitrary segments without trying to select limits coinciding with vegetation ecotones.

In 1948 Thornthwaite offered a major revision of his system, which involved more complicated calculations,[313] and yielded a much wider variety of symbol combinations. At this time he added the concept of potential evapotranspiration (i.e., the evaporative power of the air) in relation to the amount of precipitation available to meet this need. Another innovation was to recognize the capacity of the soil to store moisture for use during at least the early part of a dry season. In this system, Wapato climate is $EB'_1d(b'_2/b'_3)$.

Although this classification is quite "rational" as its author claimed, since it took advantage of the newest concepts, in actual practice the degree of fit to vegetation is not significantly closer than that proposed in 1931, or than that proposed by Köppen![104]

An Evaluation

From the phytogeographic standpoint an ideal climatic classification would yield units coinciding with vegetation Zones, i.e., areas defined on the basis of unique climatic climaxes. But no climatic classification as yet has succeeded in this, despite a constant aspiration to establish correlations with vegetation distribution. There seems to be but little doubt that each vegetation Zone has some distinctive features that set the location of its ecotones. Also it is perfectly clear that a different climatic limitation is met along different segments of the perimeter of each Zone. However, the position of each segment of an ecotone is set by competition between distinctive groups of plant species, hence the parameters that are critical may not be decisive anywhere else in the world. So long as we are considering the world as a whole, we can expect no more than a crude correlation among climatic classifications, vegetation, and soils. Maps showing the distributions of climatic climaxes appear to provide the most fundamental bioclimatic units. Zonal boundaries can be quantified, but the same parameter is not likely to be critical for more than one segment of a single Zonal boundary.[104] Vegetation is superior to soil in defining climatic units since soils come into equilibrium with climate much more slowly than vegetation approximates such an equilibrium.

A universal classification of climates that fits vegetation-defined areas better than the classifications now available is probably an unattainable aspiration.

MARINE VEGETATION

In treating the biogeography of seas and oceans a much closer integration of botany and zoology is needed than in dealing with land areas where animals are seldom dominants. In salt water animals often share dominance with plants, and in many communities they are clearly the leading dominants.

Although only major ecosystem types somewhat comparable with the Regions as recognized on land will be treated here, these are capable of subdivision as on land. Different life forms such as annuals, perennials, lime-accumulators, microscopic, macroscopic, are represented among marine plants, and community types are susceptible of definition on the basis of various combinations of plant and/or animal unions associated with specific combinations of environmental factors. As on land, ecotypic differentiation must be taken into account in assessing ecologic similarities and differences in community mosaics.

Successional relations are relatively unimportant here, except in places where bottom-dwelling plant or animal life is harvested by man.

Plankton[433]

Plankton consists mainly of microscopic algae, bacteria, fungi, and invertebrates, all of which remain suspended owing to light-weight structural materials combined with trapped gasses and oil globules. Since plankton is essentially unaffected by the depth of the water beneath the illuminated surface layer, such organisms occur throughout the seas and oceans which cover more than 70% of the earth's surface.

Except in shallow water where attached algae can get light, phytoplankton in the photic zone is the sole base of food chains, and although the density of organisms per unit volume of water averages rather low, so much area is involved that the total annual production of the oceans is believed to be approximately equal to that of the total land surface (see Table 3). The primary producers are mainly diatoms and photosynthetic types of dinoflagellates, these supporting bacteria, fungi, protozoa, and all higher forms of animal life in salt water. Near the water surface light is supraoptimal for even the algal components of plankton. Optimal conditions for their photosynthesis obtain at about 5–20 m depth depending on the clarity of the water. Zooplankton remain mostly below the phytoplankton during the day, ascending at night to graze.

Dead cells and tissues which commonly settle below the photic zone before being completely decomposed create a drain on the nutrient supply of the photic zone. Since there is no effective method for recycling nutrients back up into the photic zone, there is usually a suboptimal nutrient supply above, and an excess supply below. In tropical latitudes this is especially pronounced since the warmth of the surface layers promotes stability. This is undoubtedly a factor determining the low density of tropical plankters, despite maximum species diversity. The warmth that promotes stability also augments respiration more than photosynthesis, and this in combination with the low density of algae keeps productivity at a fairly low level. At high latitudes fertility is less limiting since surface water settles as it chills in winter, and this creates overturns which bring nutrient-enriched water to the surface. Also the photosynthesis–respiration balance is better. However, the limited supply of light energy coupled with low temperatures have a retarding influence on productivity, and the net consequence is that productivity does not vary much with latitude.

On the west sides of continents in subtropical latitudes, ocean currents cause an upwelling of nutrient-rich cold water. There plankton productivity is remarkably high, and this is reflected in the high biomass of primary and secondary consumers for which those areas are famous.

Outside the equatorial belt there is seasonal periodicity. In temperate latitudes this involves a vernal and an autumnal peak of plankton activity. At higher latitudes there is only a summer peak.

Plankton in the oceans exhibits more regular latitudinal zonation than does terrestrial vegetation. Still, near a land mass the communities are different from those of remote locations where the environment is less variable through time. Adjacent to land, temperatures differ more with the seasons, and on- and off-shore winds alter nutrient supplies by effecting overturns.

A special category of suspended vegetation occurs in the subtropical part of the central Atlantic Ocean which is called the Sargasso Sea. There two vegetatively reproducing species of brown algae, *Sargassum fluitans* and *S. natans*, remain floating by means of gas-filled bladders. The dense tangle of their floating thalli provide the framework for a unique community of other algae and animal life.

Benthon[197,352]

Benthon includes ecosystems with the dominant organisms attached to or living in close proximity to the bottom, extending as deeply as light is adequate to support attached photosynthetic plants. This environment is far more variable in time and space than is the environment of plankton, and some of the variables are as follows.

Environment

The amount of tidal action determines the length of exposure of plants growing above low tide to atmospheric drying, to freezing or heating, and to increased salinity as water films dry. In the Baltic Sea there is negligible tidal action, whereas at the opposite extreme the high and low tide levels leave wide areas exposed. In the tropics surfaces exposed at low tide have rather sparse vegetation owing to the strong insolation, unless there are breakers that provide an abundance of spray. Some intertidal algae lose so much water while the tide is out that their thalli become brittle even though they remain alive. As on land, north-facing and south-facing rock exposures may have such contrasted water and temperature relations as to support distinctly different communities.

Tropical embayments with no rivers emptying into them are typically quite saline owing to limited water mixing and high evaporation, whereas embayments fed by rivers have below-average salinity. Where fresh water trickles over the surface exposed at low tide, organisms must be attuned to diurnal alternations between freshwater and saltwater conditions—a situation in which *Enteromorpha intestinalis* (*Chlorophyta*) is especially widespread.

Where stones or gravels are small enough in relation to wave action that they are moved about, a shore can be quite barren of macroscopic life. On the other hand, immobile boulders and rock outcrops increase the amount of surface and provide dependable anchorage for attached forms of plant and animal life. Differences between limestone and granitic rocks are often reflected in the communities. Mud and sand have their own distinctive types of biotas that are quite different from each other and from rock surfaces. For example, marine vasculares (*Enalus, Halophila* and *Thalassia* in the *Hydrocharitaceae,* and *Zostera* in the *Potamogetonaceae*) nearly all require sand or mud as a substrate.

Where wave action is strong, only forms with strong holdfasts and tough thalli (Fig. 181) can form communities. Strong currents where tide levels are far apart have a similar limiting influence.

Floating ice during the spring breakup of ice packs often keep rock surfaces scoured clean except between boulders and in crevices. In large measure this accounts for the reduction in the amount of attached algae at high latitudes.

Latitudinal gradients in water temperature affect marine organisms quite markedly. Brown algae are best developed in temperate and cold latitudes, and tropical conditions have proven such an important barrier to them that the two high-latitude floras are quite distinct. Red algae, averaging smaller than browns, are best developed in tropical and subtropical waters, where

Fig. 181. Low tide on the Oregon coast has exposed *Postelsia* with its rubbery stipe raising the ribbonlike fronds about 4 dm above the rock substrate. Goose-neck barnicles (*Pollicipes polymerus*) and mussels (*Mytillus californicus*) occupy much of the area between clumps of *Postelsia*.

they share dominance with green algae. As on land, annuals become quite rare at high latitudes. Attached algae are plentiful at high latitudes even where the temperature of the water hovers about 0°C constantly, although they are not as luxuriant as in temperate latitudes. A special temperature problem for organisms growing between tide levels is that of withstanding the abrupt change when the tide returns to cover them after they have been exposed to sunlight and their tissues have become quite warm. Exposure to freezing while the tide is out can be another of life's hazards.

The effect of water depth in reducing the availability of light on the bottom is greatly augmented where there is suspended particulate matter. Many brown algae get abundant light by virtue of strong holdfasts and tough thalli which enable them to grow in shallow water where wave action and desiccation are limiting to other life forms. Other species of brown algae get sufficient light even though they grow in deep water because they have long (up to 30 m) thalli held somewhat upright by gasses trapped in special chambers. The usually more delicate red algae that are restricted to deeper water where wave action is not strong, have a compensating adaptation in the forms of a special red pigment (r phycoerythrin) which enables them to make use of light at deep levels where wavelengths are mainly in the blue region of the spectrum.

Benthic environment also includes the rock pools between tide levels, where distinctive conditions of temperature, salinity, and turbulence generate distinctive communities.

Major Divisions of Benthon

Three major worldwide divisions of benthon have been recognized, these reflecting differences in the abilities of saltwater organisms to tolerate exposure to the atmosphere.[361] In many places, secondary belts are evident within each of the following divisions.[114,213]

Supralittoral (or **Littorina–Verrucaria***) division.* Just above the limits of normal high tides there is a zone of land that is salinized by salt spray and by exceptionally high seismic or storm tides. This is the supralittoral belt, and it is usually distinguished biologically by black or at least dark-colored patches of bluegreen algae or lichens, especially *Verrucaria,* or sometimes darkly colored green or red algae, and by the presence of snails in the genus *Littorina*. Barnicles are absent.

Littoral (or **Balanus–Fucus***) division.* Surfaces regularly subject to alternate tidal submergence and emergence have their own distinctive types of communities. Here stabilized rock surfaces support dense colonies of barnicles, especially *Balanus,* locally alternating with communities of rather small brown algae, especially *Fucus,* or locally *Postelsia* (Fig. 181). *Zostera marina* (Fig. 182) is an especially important littoral dominant on both coasts of temperate North America, where by itself it may form submersed meadows that seem to stabilize muddy or sandy bottoms, or even promote deposition. Its major importance lies in providing winter food for ducks and

Fig. 182. The widely distributed rhizomatous *Zostera marina* has been exposed at low tide in the sandy bay at San Juan Island, Washington.

geese. Some fish also depend heavily on this vegetation for spawning, and many shellfish and crustaceans are components of the stands. The importance of this vegetation was not appreciated until a parasitic slime mold (*Labyrinthula*) started a drastic devastation of the vegetation in 1931. Recovery since has been hampered by the growing pollution. In the warm coastal waters around Florida, a similar monocot, *Thalassia,* forms other submarine meadows.

Infralittoral (or **Laminaria***) division.*   A rather distinctive biotic discontinuity again marks the level below which the sea bottom is but rarely and briefly exposed to air. Usually relatively large perennial algae such as *Laminaria* start to dominate the bottom at this level (Fig. 183), with the barnacles that are so characteristic of the littoral division becoming scarce. At high latitudes these algae are mostly browns, but in warm waters reds and greens tend to become more prevalent than browns. Dense growths ("forests") of these large algae provide special environments required by many free-living and epiphytic forms of life, although the plants are not very important as a direct source of food for animals other than limpets or snails. As on land, the dense stands of algae, buoyed upright, may form an overstory with a layer of shade-tolerant species beneath.[230] These algae do much to dampen wave and current action that would otherwise increase turbidity and sweep away bottom deposits of debris used by detritus feeders.

Fig. 183.   A collection of littoral brown algae with leathery thalli and holdfasts which had anchored them to the bottom below the littoral zone.

Lime-secreting algae, which supplement corals in building reefs, occur mostly in this infralittoral division. They are cosmopolitan in seas, extending to depths of 284 m, and may account for up to 90% of the material accumulating as reefs, barriers and atolls.[224,400]

Halimeda (*Chlorophyceae*), an outstanding representative of this group, deposits lime about its erect thalli, building up the reef as do coelenterates in other places, then crust-forming red algae (*Lithothamnion* or *Porolithion*) help solidify the accumulation. Although these reef-forming algae are best represented in tropical waters, they are not greatly limited by low temperatures. They are active even around Spitzbergen, and they extend into deeper water than do the coelenterate coral formers. The latter cannot grow in water colder than about 16°C, so their fossils have proven useful as criteria of past changes in the latitude of isotherms.

Reefs are highly productive of marine fauna.

Aphotic Life[238]

Autotrophes can be self-sustaining to a depth of about 200 m in the clearest waters, but beyond this plant life is represented by only bacteria. Below the level of sufficient light for photosynthesis the ocean bottom supports invertebrates and bacteria as decomposers subsisting on the rain of partially decomposed debris slowly settling from above, and these, in turn, provide the trophic base for chains of consumers, some of which are highly bizarre fish. Since the debris settles slowly, it is intercepted by detritus feeders as it descends, so very little energy-containing material reaches the bottom. Thus, the biomass there is quite limited. The ocean bottom is indeed a unique environment, with essentially constant temperature a few degrees above freezing, and with essentially no barrier to migration except pressure that varies with the topography of the ocean floor.

Land Use[441]

Considerable use has been made of littoral algae for human food, but chiefly along the Asiatic margin of the Pacific Ocean. In Japan bottoms of appropriate depth are often covered with the smallest stones that will resist turbulence, these providing the maximum surface for algae to colonize. Where the bottom is soft, sticks are stuck into the mud to provide surfaces for the attachment of desirable algae.

The larger algae are also widely harvested to extract certain gums. For example, agar and carrageenin are obtained from red algae, and algin from browns. Still other algae are harvested as important sources of iodine and potassium, which they accumulate in their metabolism. Great quantities of brown algae and *Zostera* have been gathered for use directly as fertilizer and soil conditioner in gardens.

Many grandiose but vague statements have been made regarding the possibility of depending much more on ocean life as food for man. However, current practices of man are seriously reducing the oceans' potential. Pollution has put dangerous levels of Hg and Pb in both seaweeds and associated animal life. Overharvest plus pollution have reduced the supplies of oysters and crabs. Shrimp and fish harvests are threatened if not reduced by the destruction of estuarine and mangrove vegetation.

A practical way of harvesting plankton would be to control whaling and deep sea fishing, but it is widely recognized that fish populations have been dwindling and whales are disappearing, with overharvest continuing. The anadromous fish that could annually bring thousands of tons of the highest quality protein to our very doorstep in perpetuity, are heading toward extinction in favor of hydroelectric dams of limited longevity. The Aswan dam of Egypt, by intercepting silt carrying nutrients, has eliminated an important sardine fishery about the mouth of the Nile River.

The above and other facets of the problem engender a pessimistic outlook rather than an optimistic outlook with respect to increasing the use of the seas and oceans as a source of human food.

BIBLIOGRAPHY

1. Abbe, C. (1905). A first report on the relations between climates and crops. *U.S. Dep. Agric., Weather Bur. Bull.* **36,** 386 pp.
2. Ahlgren, I. F., and Ahlgren, C. E. (1960). Ecological effects of forest fires. *Bot. Rev.* **26,** 483–533.
3. Aldon, E. F., and Springfield, H. W. (1973). The southwestern pinyon-juniper ecosystems: a bibliography. *U.S. For. Serv. Gen. Tech. Rep. RM* **4,** 20 pp.
4. Aldous, A. E. (1935). Management of Kansas permanent pastures. *Kans. Agric. Exp. Stn. Bull.* **72,** 44 pp.
5. Aleem, A. A. (1948). The recent migration of certain Indo-Pacific algae from the Red Sea into the Mediterranean. *New Phytol.* **47,** 88–94.
6. Alexander, T. R. (1953). Plant succession on Key Largo, Florida, involving *Pinus caribaea* and *Quercus virginiana Q. J. Fla. Acad. Sci.* **16,** 133–138.
7. Alexander, T. R. (1955). Observations on the ecology of the low hammocks of southern Florida. *Q. J. Fla. Acad. Sci.* **18,** 21–27.
8. Alexander, T. R. (1958). High hammock vegetation of the southern Florida mainland. *Q. J. Fla. Acad. Sci.* **21,** 293–298.
9. Allen, P. H. (1956). "The Rain Forests of Golfo Dulce," 417 pp. Univ. of Florida Press, Gainesville.
10. Anderson, E. (1937). Cytology in its relation to taxonomy. *Bot. Rev.* **3,** 335–350.
11. Anderson, E. (1949). "Introgressive Hybridization," 109 pp. Wiley, New York.
12. Antevs, E. (1932). "Alpine Zone of Mt. Washington Range," 118 pp. Meril & Webber, Auburn, Maine.
13. Arora, R. K. (1964). The forests of North Kanara District. II. Deciduous type. *Indian Bot. Soc. J.* **43,** 75–86.
14. Ashton, P. (1969). Speciation among tropical forest trees: some deductions in the light of recent evidence. *J. Ecol.* **57,** 4p–5p.
15. Aubreville, A. (1947). The disappearance of the tropical forests of Africa. *Unasylva* **1,** 5–11.
16. Axelrod, D. I. (1950). Evolution of desert vegetation in western North America. *Carnegie Inst. Washington Publ.* **590,** 215–306.
17. Axelrod, D. I. (1952). A theory of angiosperm evolution, *Evolution* **6,** 29–60.
18. Axelrod, D. I. (1958). Evolution of the Madro-Tertiary Geoflora. *Bot. Rev.* **24,** 433–509.

19. Axelrod, D. I. (1966). The Eocene Copper Basin flora of northeastern Nevada. *Univ. Calif. Publ. Geol. Sci.* **59**, 125 pp.
20. Axelrod, D. I. (1966). Origin of deciduous and evergreen habits in temperate forests. *Evolution* **20**, 1–15.
21. Bailey, R. G., and Rice, R. M. (1969). Soil slippage: an indication of instability on chaparral watersheds of southern California. *Prof. Geogr.* **21**, 172–177.
22. Baker, H. G. (1967). Support for Baker's Law—as a rule. *Evolution* **21**, 853–856.
23. Baldwin, J. J. (1968). Chaparral conversion on the Tonto National Forest. *Proc. Tall Timbers Fire Ecol. Conf.* **8**, 203–208.
24. Barbour, W. R. (1942). Forest types of tropical America. *Caribb. For.* **3**, 137–150.
25. Bartlett, H. H. (1936). A method of procedure for field work in tropical American phytogeography based on a botanical reconnaissance in parts of British Honduras and the Peten forest of Guatemala. *Carnegie Inst. Washington Publ.* **461**, 3–25.
26. Beaman, J. H. (1965). A preliminary ecological study of the alpine flora of Popocatepetl and Iztacchuatl. *Bol. Soc. Bot. Mex.* **29**, 63–74.
27. Beard, J. S. (1942). Montane vegetation in the Antilles. *Caribb. For.* **3**, 61–74.
28. Beard, J. S. (1953). The savanna vegetation of northern tropical America. *Ecol. Monogr.* **23**, 149–215.
29. Beard, J. S. (1955). The classification of tropical-American vegetation-types. *Ecology* **36**, 89–100.
30. Bentley, J. R., *et al.* (1966). Principles and techniques in converting native chaparral to stable grassland in California. *Proc. Int. Grassl. Congr.* **10**, 867–871.
31. Berry, E. W. (1925). Tertiary flora of Trinidad. *Stud. Geol.* **6**, 71–161.
32. Berry, E. W. (1930). The past climate of the North Polar region. *Smithson. Misc. Collect.* **82**, 1–29.
33. Billings, W. D. (1949). The shadscale vegetation zone of Nevada and eastern California in relation to climate and soils. *Am. Midl. Nat.* **42**, 87–109.
34. Billings, W. D. (1973). Arctic and alpine vegetations: similarities, differences, and susceptibility to disturbance. *BioScience* **23**, 697–704.
35. Billings, W. D., and Mooney, H. A. (1968). The ecology of arctic and alpine plants. *Biol. Rev. Cambridge Philos. Soc.* **43**, 481–529.
36. Bird, R. D. (1961). Ecology of the aspen parkland of western Canada in relation to land use. *Can., Dep. Agric., Publ.* **1066**, 155 pp.
37. Bisby, G. R. (1943). Geographical distribution of fungi. *Bot. Rev.* **9**, 466–482.
38. Biswell, H. H. (1956). Ecology of California grasslands. *J. Range Manage.* **9**, 19–24.
39. Blaisdell, R. S., *et al.* (1973). The role of magnolia and beech in forest processes in the Tallahassee, Florida, Thomasville, Georgia, area. *Tall Timbers Fire Ecol. Conf.* **13**, 363–397.
40. Bliss, L. C. (1956). A comparison of plant development in microenvironments of arctic and alpine tundras. *Ecol. Monogr.* **26**, 303–337.
41. Bliss, L. C. (1963). Alpine plant communities of the Presidential Range, New Hampshire, *Ecology* **44**, 678–697.
42. Bliss, L. C. (1966). Plant productivity in alpine microenvironments on Mt. Washington, New Hampshire. *Ecol. Monogr.* **36**, 125–155.
43. Bliss, L. C., and Peterson, E. B. (1973). The ecological impact of northern petroleum development. *Fond. Fran. Etud. Nord., 5th Int. Congr., Arctic Oil Gas—Probl. Possibilities.*
44. Bliss, L. C., and Woodwell, G. M. (1965). An alpine podzol on Mount Katahdin, Maine. *Soil Sci.* **100**, 274–279.
45. Bliss, L. C., *et al.* (1973). Arctic tundra ecosystems. *Annu. Rev. Ecol. Syst.* **4**, 359–399.
46. Blydenstein, J. (1968). Burning and tropical American savannas. *Proc. Tall Timbers Fire Ecol. Conf.* **8**, 1–14.

47. Böcher, T. W. (1949). The botanical expedition to west Greenland, 1946, introduction with a short mention of the vegetation areas examined. *Med. Groenl.* **147,** 1–28.
48. Boergesen, F. (1898). Notes on the shore vegetation of the Danish West Indian Islands. *Bot. Tidsskr.* **29,** 201–259.
49. Braun, E. L. (1950). "Deciduous Forest of Eastern North America," 596 pp. Blakiston, Philadelphia, Pennsylvania.
50. Briggs, J. C. (1969). The sea-level Panama Canal: potential biological catastrophe. *BioScience* **19,** 44–47.
51. Britton, M. E. (1957). Vegetation of the arctic tundra. *Proc. Annu. Biol. Colloq. Ore. State Univ.* pp. 26–61.
52. Brown, D. M. (1941). Vegetation of Roan Mountain; a phytosociological and successional study. *Ecol. Monogr.* **11,** 61–97.
53. Brown, H. E. (1958). Gambel oak in west-central Colorado. *Ecology* **39,** 317–327.
54. Brown, R. W. (1962). Paleocene flora of the Rocky Mountains and Great Plains. *U.S. Geol. Surv., Prof. Pap.* **375,** 119 pp.
55. Bryson, R. A., *et al.* (1965). Radiocarbon and soil evidence of former forest in the southern Canadian tundra. *Science* **147,** 46–48.
56. Bryson, R. A., *et al.* (1970). The character of late-glacial and post-glacial climatic changes. *in* "Pleistocene and Recent Environments of the Central Great Plains" (D. Wakefield, Jr. and J. Knox, Jr., eds.), Dep. Geol. Spec. Publ. No. 3, pp. 53–74. Univ. of Kansas Press, Lawrence.
57. Budowski, G. (1956). Tropical savannas, a sequence of forest fellings and repeated burnings. *Turrialba* **6,** 23–33.
58. Burcham, L. T. (1957). "California Range Land," 261 pp.
59. Burzlaff, D. F. (1962). A soil and vegetation inventory and analysis of three Nebraska sandhills range sites. *Neb., Agric. Exp. Stn., Res. Bull.* **206,** 32 pp.
60. Cain, S. A. (1933). An ecological study of the heath balds of the Great Smoky Mountains of Tennessee and North Carolina. *Butler Univ. Bot. Stud.* **1,** 177–208.
61. Camp, W. H. (1951). Biosystematy. *Brittonia* **7,** 113–127.
62. Campbell, R. S. (1929). Vegetative succession in the *Prosopis* sand dunes of southern New Mexico. *Ecology* **10,** 392–398.
63. Carlquist, S. (1967). The biota of long-distance dispersal. V. Dispersal to Pacific Islands. *Bull. Torrey Bot. Club* **94,** 129–162.
64. Carlson, M. C. (1954). Floral elements of the pine-oak-*Liquidambar* forest of Montebello, Chiapas, Mexico. *Bull. Torrey Bot. Club.* **81,** 387–400.
65. Carr, A. F., Jr. (1950). Outline for a classification of animal habitats in Honduras. *Bull. Am. Mus. Nat. Hist.* **94,** 563–594.
66. Chabot, B. F., and Billings, W. D. (1972). Origins and ecology of the Sierra alpine flora and vegetation. *Ecol. Monogr.* **42,** 163–199.
67. Chaney, R. W., and Axelrod, D. I. Miocene floras of the Columbia Plateau. *Carnegie Inst. Washington Publ.* **617,** 237 pp.
68. Chapman, V. J. (1946). Marine algal ecology. *Bot. Rev.* **12,** 628–672.
69. Chapman, V. J. (1970). Mangrove phytosociology. *Trop. Ecol.* **11,** 1–19.
70. Christensen, E. M. (1955). Ecological notes on the mountain brush in Utah. *Proc. Utah Acad. Sci., Arts, Lett.* **32,** 107–111.
71. Christensen, E. M. (1959). A comparative study of mountain brush, pinyon-junipeer, and sagebrush communities in Utah. *Proc. Utah Acad. Sci., Arts, Lett.* **36,** 174–175.
72. Clark, H. W. (1937). Association types of the north Coast Ranges of California. *Ecology* **18,** 214–231.
73. Clarke, S. E., *et al.* (1943). The effects of climate and grazing on shortgrass prairie vegetation. *Can., Dep. Agric., Tech. Bull.* **46,** 53 pp.

74. Clary, W. P., *et al.* (1974). Effects of pinyon-juniper removal on natural resource products and uses in Arizona. *U.S. For. Serv. Res. Pap. RM* **128**, 28 pp.

75. Clements, F. E. (1918). Scope and significance of paleo-ecology. *Geol. Soc. Am. Bull.* **29**, 369–374.

76. Cline, A. C., and Spurr, S. H. (1942). The virgin upland forest of central New England. *Bull. Harvard For.* **21**, 1–58.

77. Conard, H. S. (1938). The fir forests of Iowa. *Proc. Iowa Acad. Sci.* **45**, 69–72.

78. Cooper, W. S. (1936). The strand and dune flora of the Pacific coast of North America. *In* "Essays in Geobotany in Honor of William Albert Setchell" (T. H. Goodspeed, ed.), pp. 141–187. Univ. of California Press, Berkeley.

79. Corbet, P. S. (1969). Terrestrial microclimate: amelioration at high latitudes. *Science* **166**, 865–866.

80. Corte, A. (1970). Bioecological aspects of the snow plant communities of Cape Spring, Argentine Antarctica. *In* "Ecology of the Subarctic Regions," pp. 101–104. UNESCO, New York.

81. Costin, A. B. (1965). Long-distance seed dispersal to Macquarie Island. *Nature (London)* **206**, 317.

82. Cottam, W. P., *et al.* (1959). Some clues to Great Basin postpluvial climates provided by oak distribution. *Ecology* **40**, 361–377.

83. Cottle, H. J. (1931). Studies in the vegetation of southwestern Texas. *Ecology* **12**, 105–155.

84. Coupland, R. T. (1950). Ecology of mixed prairie in Canada. *Ecol. Monogr.* **20**, 271–315.

85. Coupland, R. T. (1961). A reconsideration of grassland classification in the northern Great Plains of North America. *J. Ecol.* **49**, 135–167.

86. Cox, C. F. (1933). Alpine plant succession on James Peak, Colorado, *Ecol. Monogr.* **3**, 299–372.

87. Crandall, D. L. (1958). Ground vegetation patterns of the spruce-fir area of the Great Smoky Mountains National Park. *Ecol. Monogr.* **28**, 337–360.

88. Cruden, R. W. (1966). Birds as agents of long-distance dispersal for disjunct plant groups of the western hemisphere. *Ecology* **20**, 517–563.

89. Cuatrecasas, J. (1957). A sketch of the vegetation of the north-Andean province. *Pac. Sci. Congr. Proc. 8th* **4**, 167–173.

90. Cuatrecasas, J. (1968). Paramo vegetation and its life forms. *Colloq. Geogr.* **9**, 163–186.

91. Cutler, D. F. (1972). Vicarious species of Restionaceae in Africa, Australia and South America. *In* "Taxonomy, Phytogeography and Evolution" (D. H. Valentine, ed.), pp. 73–83. Academic Press, New York.

92. Cypert, E. (1973). Plant succession on burned areas in Okefinokee Swamp following the fires of 1954 and 1955. *Proc. Tall Timbers Fire Ecol. Conf. Anim.* **12**, 199–217.

93. Dahl, E. (1956). Rondane Mountain vegetation in south Norway and its relation to the environment. *Skr. Nor. Vidensk. Akad. Oslo, 1* **3**, 373 pp.

94. Dansereau, P. (1954). Studies on central Baffin vegetation. 1. Bray Island. *Vegetatio* **6**, 329–339.

95. Dansereau, P. (1959). Phytogeographia laurentiana. II. The principal plant associations of the St. Lawrence Valley. *Contrib. Inst. Bot. Univ. Montreal* **75**, 147 pp.

96. Darlington, C. D. (1956). "Chromosome Botany," 186 pp. Allen & Unwin, London.

97. Daubenmire, R. (1936). The "Big Woods" of Minnesota: Its structure, and relation to climate, fire and soils. *Ecol. Monogr.* **6**, 233–268.

98. Daubenmire, R. (1943). Vegetational zonation in the Rocky Mountains. *Bot. Rev.* **9**, 325–393.

99. Daubenmire, R. (1953). Notes on the vegetation of forested regions of the far northern Rockies and Alaska. *Northwest Sci.* **27**, 125–137.

100. Daubenmire, R. (1954). Alpine timberlines in the Americas and their interpretation. *Butler Univ. Bot. Stud.* **11,** 119–136.

101. Daubenmire, R. (1957). Injury to plants from rapidly dropping temperatures in Washington and northern Idaho. *J. For.* **55,** 581–585.

102. Daubenmire, R. (1968). Soil moisture in relation to vegetation distribution in the mountains of northern Idaho. *Ecology* **49,** 431–438.

103. Daubenmire, R. (1969). Annual cycles of soil moisture and temperature as related to grass development in the steppe of eastern Washington. *Ecology* **53,** 419–424.

104. Daubenmire, R. (1970). Steppe vegetation of Washington. *Wash., Agric. Exp. Stn., Tech. Bull.* **62,** 131 pp.

105. Daubenmire, R. (1972). Some ecological consequences of converting forest to savanna in northwestern Costa Rica. *Trop. Ecol.* **13,** 31–51.

106. Daubenmire, R. (1972). Phenology and other characteristics of tropical semi-deciduous forest in northwestern Costa Rica. *J. Ecol.* **60,** 147–170.

107. Daubenmire, R. (1974). Taxonomic and ecologic relationships between *Picea glauca* and *P. engelmannii. Can. J. Bot.* **52,** 1545–1560.

108. Daubenmire, R. (1975). Floristic plant geography of eastern Washington and northern Idaho. *J. Biogeog.* **2,** 1–18.

109. Daubenmire, R., and Daubenmire, J. B. (1968). Forest vegetation of eastern Washington and northern Idaho. *Wash., Agric. Exp. Stn., Tech. Bull.* **60,** 104 pp.

110. Davis, J. H., Jr. (1940). The ecology and geologic role of mangroves in Florida. *Carnegie Inst. Washington Publ.* **517,** 303–412.

111. Davis, R. B. (1966). Spruce-fir forests of the coast of Maine. *Ecol. Monogr.* **36,** 79–94.

112. Deevey, E. S. (1949). Biogeography of the Pleistocene. *Geol. Soc. Am. Bull.* **60,** 1315–1416.

113. Delcourt, H. R., and Delcourt, P. A. (1974). Primeval magnolia-holly-beech climax in Louisiana. *Ecology* **55,** 638–644.

114. Detling, L. E. (1961). The chaparral formation of southwestern Oregon, with considerations of its postglacial history. *Ecology* **42,** 348–357.

115. Dietz, R. S., and Holden, J. C. (1970). The breakup of Pangaea. *Sci. Am.* **223**(4), 30–41.

116. Donley, D. E., and Mitchell, R. L. (1939). The relation of rainfall to elevation in the southern Appalachian region. *Trans., Am. Geophys. Union* pp. 711–721.

117. Dorf, E. (1960). Climatic changes of past and present. *Science* **48,** 341–364.

118. Doty, M. S. (1946). Critical tide factors that are correlated with the vertical distribution of marine algae and other organisms along the Pacific coast. *Ecology* **27,** 315–328.

119. Doyle, J. A. (1969). Cretaceous angiosperm pollen of the Atlantic coastal plain and its evolutionary significance. *J. Arnold Arbor, Harvard Univ.* **50,** 1–35.

120. Dressler, R. L. (1954). Some floristic relationship between Mexico and the United States, *Rhodora* **56,** 81–95.

121. Duke, J. A., and Terrell, E. E. (1974). Crop diversification matrix: Introduction. *Taxon* **23,** 759–799.

122. Dyksterhuis, E. J. (1948). The vegetation of the western Cross Timbers. *Ecol. Monogr.* **18,** 325–376.

123. Egler, F. E. (1950). Southeast saline everglades vegetation, Florida and its management. *Vegetatio* 3, 213–265.

124. Eiten, G. (1972). The cerrado vegetation of Brazil. *Bot. Rev.* **38,** 201–341.

125. Elias, M. K. (1942). Tertiary prairie grasses and other herbs from the High Plains. *Geol. Soc. Am., Spec. Paper.* **41,** 176 pp.

126. Ellison, L. (1954). Subalpine vegetation of the Wasatch Plateau, Utah. *Ecol. Monogr.* **24,** 89–184.

127. Emerson, F. W. (1935). An ecological reconnaissance in the White Sands, New Mexico. *Ecology* **16,** 226–233.

128. Emiliani, C. (1972). Quaternary paleotemperatures and the duration of the high-temperature intervals. *Science* **178**, 398–401.
129. Epling, C., and Lewis, H. (1942). The centers of distribution of the chaparral and coastal sage associations. *Am. Midl. Nat.* **27**, 445–462.
130. Ewing, J. (1924). Plant successions of the brush-prairie in northwestern Minnesota. *J. Ecol.* **12**, 238–266.
131. Eyre, S. R. (1968). "Vegetation and Soils, a World Picture," 2nd Ed., 314 pp. Arnold, London.
132. Faegri, K. (1963). Problems of immigration and dispersal of the Scandinavian flora. *In* "North Atlantic Biota and their History" (A. Löve and D. Löve, eds.), pp. 221–232. Macmillan, New York.
133. Farrar, D. R. (1967). Gametophytes of four tropical fern genera reproducing independently of their sporophytes in the southern Appalachians. *Science* **155**, 1266–1267.
134. Fassett, N. C. (1943). Another driftless area endemic. *Bull. Torrey Bot. Club* **70**, 398–399.
135. Fautin, R. W. (1946). Biotic communities of the northern desert scrub biome in western Utah. *Ecol. Monogr.* **16**, 251–310.
136. Fernald, M. L. (1925). Persistence of plants in unglaciated areas of boreal America. *Mem. Am. Acad. Arts Sci.* **15**, 239–242.
137. Fischer, A. G. (1960). Latitudinal variation in organic diversity. *Evolution* **14**, 64–81.
138. Flint, R. F., and Deevey, E. S., Jr. (1951). Radiocarbon dating of late-Pleistocene events. *Am. J. Sci.* **249**, 257–300.
139. Flint, R. F., *et al.* (1942). Glaciation of Shickshock Mountains, Gaspe Peninsula. *Geol. Soc. Am. Bull.* **53**, 1211–1230.
140. Flowers, S. (1934). Vegetation of the Great Salt Lake region. *Bot. Gaz.* **95**, 353–418.
141. Fosberg, F. R. (1938). The Lower Sonoran in Utah. *Science* **87**, 39–40.
142. Fosberg, F. R. (1944). El páramo de Sumapaz, Colombia. *N.Y. Bot. Gard. J.* **45**, 226–234.
143. Fosberg, F. R. (1948). Derivation of the flora of the Hawaiian Islands. *In* "Insects of Hawaii" (E. C. Zimmerman, ed.), Vol. 1, pp. 107–119. Univ. of Hawaii Press, Honolulu.
144. Fowells, H. A., ed. (1900). Silvics of forest trees of the United States. *U.S. Dep. Agric., Agric. Handb.* **271**, 762 pp.
145. Franklin, J. F., and Dyrness, C. T. (1973). Natural vegetation of Oregon and Washington, *U.S. For. Serv. Tech. Rep. PNW* **8**, 417 pp.
146. Franklin, J. F., *et al.* (1971). Invasion of subalpine meadows by trees in the Cascade Range, Washington and Oregon. *Arctic Alpine Res.* **3**, 215–224.
147. Frenzel, B. (1968). The Pleistocene vegetation of northern Eurasia. *Science* **161**, 637–649.
148. Froiland, S. G. (1952). The biological status of *Betula andrewsii* A. Nels. *Evolution* **6**, 268–282.
149. Gallegly, M. E., and Galindo, J. (1958). Mating types and oospores of *Phytophthora infestans* in nature in Mexico. *Phytopathology* **48**, 274–277.
150. Garren, K. H. (1943). Effects of fire on vegetation of the southeastern United States. *Bot. Rev.* **9**, 617–654.
151. Gates, F. C. (1942). Bogs of northern Lower Michigan. *Ecol. Monogr.* **12**, 213–254.
152. Gay, P. A. (1960). The water hyacinth and the Sudan. *In* "The Biology of Weeds" (J. L. Harper, ed.), pp. 184–188. Blackwell, Oxford.
153. Gentry, H. S. (1942). A study of the flora and vegetation of the valley of the Rio Maya, Sonora. *Carnegie Inst. Washington Publ.* **527**, 328 pp.
154. Gjaerevoll, O. (1956). The plant communities of the Scandinavian alpine snow-beds. *K. Nor. Vidensk. Selsk. Skr.* No. 1, 1–405.
155. Gjaerevoll, O. (1963). Survival of plants on nunataks in Norway during the Pleistocene glaciation. *In* "North Atlantic Biota and their History" (A. Löve and D. Löve, eds.), pp. 261–284. Macmillan, New York.

156. Gleason, H. A. (1924). Age and area from the viewpoint of phytogeography. *Am. J. Bot.* 11, 541–546.
157. Gomez-Pompa, A. (1973). Ecology of the vegetation of Vera Cruz. *In* "Vegetation and Vegetational History of Northern Latin America" (A. Graham, ed.), pp. 73–148. Elsevier, Amsterdam.
158. Gould, F. W. (1968). "Grass Systematics," 382 pp. McGraw-Hill, New York.
159. Graham, A. (1972). Outline of the origin and historical recognition of floristic affinities between Asia and eastern North America. *In* "Flo159. Graham, A. (1972). Outline of the origin and historical recognition of floristic affinities between Asia and eastern North between Asia and eastern North America. *In* "Floristics and Paleofloristics of Asia and Eastern North America" (A. Graham, ed.), pp. 1–18. Elsevier, Amsterdam.
160. Graham, A. (1973). History of the arborescent temperate element in the northern Latin American biota. *In* "Vegetation and Vegetational History of Northern Latin America" (A. Graham, ed.), pp. 301–314. Elsevier, Amsterdam.
161. Grant, V. (1971). "Plant Speciation," 435 pp. Columbia Univ. Press, New York.
162. Griggs, R. F. (1934). The edge of the forest in Alaska and the reasons for its position. *Ecology* 15, 80–96.
163. Griggs, R. F. (1942). Indications as to climatic changes from the timberline of Mt. Washington. *Science* 95, 515–519.
164. Griggs, R. F. (1946). The timberlines of northern America and their interpretation. *Ecology* 27, 275–289.
165. Grubb, P. J. (1971). Interpretation of the "massenerhebung" effect on tropical mountains. *Nature (London)* 229, 44–45.
166. Grubb, P. J., and Whitmore, T. C. (1966). A comparison of montane and lowland rain forest in Ecuador. II. The climate and its effects on the distribution and physiognomy of the forests. *J. Ecol.* 54, 303–333.
167. Grüger, E. (1972). Pollen and seed studies of Wisconsinan vegetation in Illinois, U.S. *Geol. Soc. Am. Bull.* 83, 2715–2734.
168. Habeck, J. R. (1958). White cedar ecotypes in Wisconsin. *Ecology* 39, 457–463.
169. Hadac, E. (1963). On the history and age of some Arctic plant species. *In* "North Atlantic Biota and their History" (A. Löve and D. Löve, eds.), pp. 207–220. Macmillan, New York.
170. Hallam, A. (1967). The bearing of certain paleozoogeographic data on continental drift. *Palaeogeogr., Palaeoclimatol., Palaeoecol.* 3, 201–241.
171. Hamilton, E. L. (1956). Sunken islands of the mid-Pacific mountains. *Geol. Soc. Am., Mem.* 64, 97 pp.
172. Hanes, T. L. (1971). Succession after fire in the chaparral of southern California. *Ecol. Monogr.* 41, 27–52.
173. Hanson, H. C. (1953). Vegetation types in northwestern Alaska and comparisons with communities in other Arctic regions. *Ecology* 34, 111–140.
174. Hanson, H. C., and Whitman, W. (1938). Characteristics of major grasslands types in western North Dakota. *Ecol. Monogr.* 8, 57–114.
175. Hara, H. (1972). Patterns of differentiation in flowering plants. *In* "Floristics and Paleofloristics of Asia and Eastern North America" (A. Graham, ed.), pp. 55–60. Elsevier, Amsterdam.
176. Hare, F. K. (1951). Climatic classification. *In* "London Essays in Geography" (L. D. Stamp, ed.), pp. 111–134. Harvard Univ. Press, Cambridge, Massachusetts.
177. Hare, F. K. (1961). The causation of the arid zone. *Arid Zone Res.* 17, 25–30.
178. Harper, R. M. (1911). The relation of climax vegetation to islands and peninsulas. *Bull. Torrey Bot. Club* 38, 515–525.
179. Harris, J. A. (1917). Physical chemistry in the service of phytogeography. *Science* 46, 25–30.

180. Harrison, A. T., *et al.* (1971). Drought relationships and distribution of two Mediterranean-climate California plant communities. *Ecology* **52**, 869–875.

181. Hastings, J. R., and Turner, R. M. (1965). "The Changing Mile," 317 pp. Univ. of Arizona Press, Tucson.

182. Hawksworth, F. G., and Wiens, D. (1972). Biology and classification of dwarf mistletoes (*Arceuthobium*). *U.S. Dep. Agric., Agric. Handb.* **401**, 234 pp.

183. Hay, O. P. (1923). The Pleistocene of North America and its vertebrated animals from the states east of the Mississippi River and from the Canadian provinces east of longitude 95°. *Carnegie Inst. Washington Publ.* **322**, 499 pp.

184. Hay, O. P. (1924). The Pleistocene of the middle region of North America and its vertebrated animals. *Carnegie Inst. Washington Publ.* **322A**, 385 pp.

185. Hedberg, O. (1964). "Features of Afroalpine Plant Ecology," 144 pp. Almqvist & Wiksell, Stockholm.

186. Hedgpeth, J. W. (1957). Treatise on marine ecology and paleoecology. Vol. 1. Ecology. *Geol. Soc. Am., Mem.* **67**, 1296 pp.

187. Heezen, B. C., and Tharp, M. (1963). The Atlantic floor. *In* "North Atlantic Biota and their History" (A. Löve and D. Löve, eds.), pp. 21–27. Macmillan, New York.

188. Hellmers, H., *et al.* (1955). Root systems of some chaparral plants in southern California. *Ecology* **36**, 667–678.

189. Henwood, K. (1973). A structural model of forces in buttressed tropical rainforest trees. *Biotropica* **5**, 83–89.

190. Heslop-Harrison, J. (1964). Forty years of genecology. *Adv. Ecol. Res.* **2**, 159–264.

191. Heyligers, P. C. (1963). Vegetation and soil of a white-sand savanna in Suriname. *Verh. K. Ned. Akad. Wet., Afd. Natuurk., Reeks* 00 54(3), 1–148.

192. Hodge, W. H. (1946). Cushion plants of the Peruvian puna. *N.Y. Bot. Gard. J.* **47**, 133–141.

193. Hoff, C. C. (1957). A comparison of soil, climate, and biota of conifer and aspen communities in the central Rocky Mountains. *Am. Midl. Nat.* **58**, 115–140.

194. Hoffman, G. R., and Kazmierski, R. G. (1969). An ecological study of epiphytic bryophytes on *Pseudotsuga menziesii* on the Olympic Peninsula, Washington. 1. A description of the vegetation. *Bryologist* **72**, 1–19.

195. Hoffman, G. R., and Timken, R. L. (1970). Ecological observations on *Pinus ponderosa* Laws. (*Pinaceae*) at its easternmost extension in South Dakota. *Southwest Nat.* **14**, 327–336.

196. Holdgate, M. W. ed. (1970). "Antarctic Ecology," 2 Vols. Academic Press, New York

197. Holme, N. A., and McIntyre, A. D., eds. (1971). "Methods for the Study of Marine Benthos," I.B.P. Handbook No. 16, 336 pp. Davis, Philadelphia, Pennsylvania.

198. Holmes, C. H. (1951). "The Grass, Fern, and Savannah Lands of Ceylon, their Nature and Ecological Significance," Imp. For. Inst. Pap. No. 28, 95 pp. Oxford Univ. Press, London and New York.

199. Holttum, R. E. (1931). On periodic leaf-change and flowering of trees in Singapore. *Gard. Bull. (Singapore)* **5**, 173–206.

200. Holmes, C. H. (1939–1940). The uniform climate of Malaya as a barrier to plant migration. *Pac. Sci. Congr. Proc., 6th* **4**, 669–671.

201. Horton, J. S., and Kraebel, C. J. (1955). Development of vegetation after fire in the chamise chaparral of southern California. *Ecology* **36**, 244–262.

202. Howden, H. F. (1974). Problems in interpreting dispersal of terrestrial organisms as related to continental drift. *Biotropica* **6**, 1–6.

203. Hultén, E. (1937). "Outline of the History of Arctic and Boreal Biota during the Quaternaru Period," 168 pp. Bokfoer. Aktieb. Thule, Stockholm.

204. Humphrey, R. R. (1958). The desert grassland, a history of vegetational change and an analysis of causes. *Bot. Rev.* **24**, 193–252.

205. Hurley, D. P. (1968). The confirmation of continental drift. *Sci. Am.* (April), 52–64.
206. Husted, W. M. (1970). Altithermal occupation of the northern Rocky Mountains by plains hunting peoples. *Am. Quat. Assoc., 1st Meet.* Abstr. p. 69.
207. Hustich, I. (1948). The Scotch pine in northernmost Finland and its dependence on the climate in the last decades. *Acta Bot. Fenn.* **42,** 75 pp.
208. Hustich, I. (1949). On the forest geography of the Labrador Peninsula. A preliminary synthesis. *Acta Geogr.* **10**(2), 1–63.
209. Hustich, I. (1953). The boreal limits of conifers. *Arctic* **6,** 149–162.
210. Hustich, I. (1962). A comparison of the floras on subarctic mountains in Labrador and in Finnish Lapland. *Acta Geogr.* **17,** 1–24.
211. Ives, R. A. (1942). Atypical subalpine environments. *Ecology* **23,** 89–96.
212. Jameson, D. A. (1962). Effects of burning on a galleta-black grama range invaded by juniper. *Ecology* **43,** 760–763.
213. Janzen, D. H. (1973). Tropical agroecosystems. *Science* **182,** 1212–1219.
214. Jardine, N., and McKenzie, D. (1972). Continental drift and the evolution of organisms. *Nature (London)* **235,** 20–24.
215. Jenny, H. (1948). Great Soil Groups in the equatorial regions of Colombia, South America. *Soil Sci.* **66,** 5–28.
216. Jenny, H., *et al.* (1969). The pygmy forest-podzol ecosystem and its dune associates of the Mendocino coast. *Madrono* **20,** 60–74.
217. Jewitt, T. N. (1966). Soils of arid lands. *In* "Arid Lands" (E. S. Hills, ed.), pp. 103–125. Methuen, London.
218. Johnson, A. M. (1927). The bitternut hickory, *Carya cordiformis*, in northern Minnesota. *Am. J. Bot.* **14,** 49–51.
219. Johnson, A. W. (1968). The evolution of desert vegetation in western North America. *In* "Desert Biology" (G. W. Brown, ed.), Vol. 1, pp. 101–140. Academic Press, New York.
220. Johnson, A. M., and Packer, J. G. (1966). Distribution, ecology and cytology of the Ogotoruk Creek flora and the history of Beringia. *In* "The Bering Land Bridge" (D. M. Hopkins, ed.), pp. 245–265. Stanford Univ. Press, Stanford, California.
221. Johnson, A. M., *et al.* (1965). Polyploidy, distribution, and environment. *In* "The Quaternary of the United States" (H. E. Wright and D. G. Frey, eds.), pp. 497–507. Princeton Univ. Press, Princeton, New Jersey.
222. Johnson, A. M., *et al.* (1966). Vegetation and flora of the Cape Thompson-Ogortoruk Creek area, Alaska. *In* "Environment of the Cape Thompson Region Alaska" (N. J. Wilimovsky and J. N. Wolfe, eds.), pp. 277–354. US AEC, Washington, D.C.
223. Johnson, D. S., and Skutch, A. F. (1928). Littoral vegetation on a headland of Mt. Desert Island, Maine. *Ecology* **9,** 188–215, 307–338.
224. Johnson, J. H. (1936). Algae as rock builders, with notes on some algal limestones from Colorado. *Univ. Colo. Stud.* **23,** 217–222.
225. Johnson, P. L. (1969). Arctic plants, ecosystems and strategies. *Arctic* **22,** 341–355.
226. Johnson, P. L., and Billings, W. D. (1962). The alpine vegetation of the Beartooth Plateau in relation to crypedogenic processes and patterns. *Ecol. Monogr.* **32,** 105–135.
227. Johnston, M. C. (1963). Past and present grasslands of southern Texas and northeastern Mexico. *Ecology* **44,** 456–466.
228. Just, T. (1952). Fossil floras of the southern hemisphere and their phytogeographical significance. *Bull. Am. Mus. Nat. Hist.* **99,** 189–203.
229. Kimball, H. H. (1928). Amount of solar radiation that reaches the surface of the earth on the land and on the sea, and methods by which it is measured. *Mon. Weather Rev.* **56,** 393–399.
230. Kitching, J. A. (1941). Studies in sublittoral ecology. III. Laminaria forest on the west coast of Scotland; a study of zonation in relation to wave action. *Biol. Bull.* **180,** 324–337.

231. Kittredge, J. (1955). Litter and forest floor of the chaparral in parts of the San Dimas Experimental Forest, California. *Hilgardia* 23, 563–596.

232. Koch, B. E. (1964). Review of fossil floras and nonmarine deposits of west Greenland. *Geol. Soc. Am. Bull.* 75, 535–548.

233. Koriba, K. (1958). On the periodicity of tree-growth in the tropics, with reference to the mode of branching, the leaf-fall, and the formation of the resting bud. *Gard. Bull. (Singapore)* 17, 11–81.

234. Kornas, J. (1972). Corresponding taxa and their ecological background in the forests of temperate Eurasia and North America. *In* "Taxonomy, Phytogeography and Evolution" (D. H. Valentine, ed.), pp. 37–59. Academic Press, New York.

235. Korstian. C. F. (1937). Perpetuation of spruce on cut-over and burned lands in the higher southern Appalachian Mountains. *Ecol. Monogr.* 28, 293–336.

236. Krajina, V. J., *et al.* (1963). Ecology of the forests of the Pacific Northwest. *Univ. B.C., Dep. Biol. Bot., Prog. Rep. 1962*, 104 pp.

237. Krammes, J. S. (1965). Seasonal debris movement from steep mountainside slopes in southern California. *U.S. Dep. Agric., Misc. Publ.* 970, 85–88.

238. Krogh, A. (1934). Conditions of life at great depths in the ocean. *Ecol. Monogr.* 4, 430–439.

239. Kucera, C. L., and McDermott, R. E. (1955). Sugar maple-basswood studies in the forest-prairie transition of central Missouri. *Am. Midl. Nat.* 54, 495–503.

240. Kucera, C. L., and Martin, S. C. (1957). Vegetation and soil relationships in the glade region of the southwestern Missouri Ozarks. *Ecology* 38, 285–291.

241. Küchler, A. W. (1946). The broadleaf deciduous forests of the Pacific Northwest. *Assoc. Am. Geogr., Ann.* 36, 122–147.

242. Kuchler, A. W. (1964). Potential natural vegetation of the conterminous United States. *Am. Geogr. Soc., Spec. Publ.* 36, 155 pp. (With large wall map.)

243. Kurz, H. (1933). Northern disjuncts in northern Florida. *Fla. Geol. Surv., Annu. Rep.* 23/24:50–53.

244. Kurz, H. (1945). Secondary forest succession in the Tallahassee Red Hills. *Proc. Fla. Acad. Sci.* 7, 59–100.

245. Laessle, A. M. (1942). The plant communities of the Welaka area, with special reference to correlations between soils and vegetational succession. *Univ. Fla. Publ., Biol. Sci. Ser.* 4, 143 pp.

246. Laessler, A. M. (1958). The origin and successional relationship of sandhill vegetation and sand-pine scrub. *Ecol. Monogr.* 28, 361–387.

247. Laessler, A. M. (1968). Relationship of sand pine scrub to former shore lines. *Q. J. Fla. Acad. Sci.* 30, 269–286.

248. La Marche, V. C., Jr., and Mooney, H. A. (1967). Altithermal timberline advance in western United States. *Nature (London)* 213, 980–982.

249. LaRoi, G. H. (1967). Ecological studies in the boreal spruce-fir forests of the North American taiga. 1. Analysis of the vascular flora. *Ecol. Monogr.* 37, 229–253.

250. Lauer, W. (1968). (Problems in the phytogeographic division of Central America). *Colloq. Geogr.* 9, 139–156.

251. Leighly, J. (1949). Climatology since the year 1800. *Trans., Am. Geogr. Union* 30, 658–672.

252. Leopold, A. S. (1950). Vegetation zones of Mexico. *Ecology* 31, 507–518.

253. Leopold, E. B. (1967). Late-Cenozoic patterns of plant extinction. *In* "Pleistocene Extinctions, the Search for a Cause" (P. S. Martin and H. E. Wright, Jr., eds.), pp. 203–246. Yale Univ. Press, New Haven, Connecticut.

254. Leopold, E. B., and MacGinitie, H. D. (1972). Development and affinities of Tertiary floras in the Rocky Mountains. *In* "Floristics and Paleofloristics of Asia and Eastern North America" (A. Graham, ed.), pp. 147–200. Elsevier, New York.

255. LeSueur, H. (1945). The ecology of the vegetation of Chihuahua, Mexico north of parallel 28. *Univ. Tex. Bull.* **4521,** 92 pp.
256. Li, H.-L. (1952). Floristic relationships between eastern Asia and eastern North America. *Trans. Am. Philos. Soc.* **42,** 371–429.
257. Lieth, H. (1972). Modelling the primary productivity of the world. *Trop. Ecol.* **13,** 125–130.
258. Lima, D. de A. (1965). Vegetation of Brazil. *Proc. Int. Grassl. Congr., 9th* **1,** 29–47.
259. Livingstone, D. A. (1955). Some pollen profiles from Arctic Alaska. *Ecology* **36,** 587–600.
260. Löve, A. (1959). Origin of the Arctic flora. *In* "Problems of the Pleistocene and Arctic," Mus. Publ. No. 1, pp. 82–95. McGill Univ. Press, Montreal, Canada.
261. Löve, D. (1963). Dispersal and survival of plants. *In* "North Atlantic Biota and their History" (A. Löve and D. Löve, eds.), pp. 189–205. Macmillan, New York.
262. Loveless, C. M. (1959). A study of the vegetation in the Florida everglades. *Ecology* **40,** 1–9.
263. Lugo, A. E., and Snedaker, S. C. (1974). The ecology of mangroves. *Annu. Rev. Ecol. Syst.* **5,** 39–64.
264. Lundell, C. L. (1937). The vegetation of Petèn. *Carnegie Inst. Washington Publ.* **478,** 244 pp.
265. Lutz, H. J. (1930). Original forest composition in northwestern Pennsylvania as indicated by early land survey notes. *J. For.* **28,** 1098–1103.
266. Lynch, D. (1955). Ecology of the aspen groveland in Glacier County, Montana. *Ecol. Monogr.* **25,** 321–344.
267. McAvoy, B. (1931). Ecological survey of the Bella Coola region. *Bot. Gaz.* **92,** 141–171.
268. McCormick, J., and Buell, M. F. (1968). The plains: pygmy forests of the New Jersey pine barrens, a review and annotated bibliography. *N.J. Acad. Sci. Bull.* **13,** 20–34.
269. MacGinitie, H. D. (1933). Redwoods and frost. *Science* **78,** 190.
270. MacGinitie, H. D. (1953). Fossil plants of the Florissant beds, Colorado. *Carnegie Inst. Washington Publ.* **599,** 198 pp.
271. MacGinitie, H. D. (1959). Climate since the late Cretaceous. *In* "Zoogeography, a Symposium" (C. L. Hubbs, ed.), pp. 61–79. A.A.A.S Publ. No. 51, Washington, D.C.
272. McLean, A. (1970). Plant communities of the Similkameen Valley, British Columbia, and their relationship to soils. *Ecol. Monogr.* **40,** 403–424.
273. McMillan, C. (1956). The edaphic restriction of *Cupresus* and *Pinus* in the Coast Ranges of central California. *Ecol. Monogr.* **26,** 177–212.
274. McPherson, J. K., and Muller, C. H. (1969). Allelopathic effects of *Adenostoma fasciculatum,* "chamise," in the California chaparral. *Ecol. Monogr.* **39,** 177–198.
275. Mark, A. F. (1958). The ecology of the southern Appalachian grass balds. *Ecol. Monogr.* **28,** 293–336.
276. Marr, J. W. (1948). Ecology of the forest-tundra ecotone on the east coast of Hudson Bay. *Ecol. Monogr.* **18,** 11–144.
277. Martin, P. S. (1958). A biogeography of reptiles and amphibians in the Gomez Farias region, Tamaulipas, Mexico. *Univ. Mich. Mus. Zool., Misc. Publ.* **101,** 102 pp.
278. Martin, P. S. (1958). Taiga-tundra and the full-glacial period in Chester County, Pennsylvania. *Am. J. Sci.* **256,** 470–502.
279. Martin, P. S., and Gray, J. (1962). Pollen analysis and the Cenozoic. *Science* **137,** 103–111.
280. Martin, P. S., and Harrell, B. E. (1957). The Pleistocene history of temperate biotas in Mexico and eastern United States. *Ecology* **38,** 470–480.
281. Martin, P. S., and Mehringer, P. J. (1965). Pleistocene pollen analysis and biogeography of the southwest. *In* "The Quaternary of the United States" (H. E. Wright, Jr., and D. G. Frey, eds.), pp. 433–452. Princeton Univ. Press, Princeton, New Jersey.

282. Marvin, U. B. (1973). "Continental Drift, the Evolution of a Concept," 239 pp. Random House (Smithsonian Inst. Press), New York.
283. Maxwell, J. A., and Davis, M. B. (1972). Pollen evidence of Pleistocene and Holocene vegetation on the Allegheny Plateau, Maryland. *Quat. Res. (N.Y.)* **1**, 506–530.
284. Mayr, E., ed. (1952). The problem of land connections across the South Atlantic, with special reference to the Mesozoic. *Bull. Am. Mus. Nat. Hist.* **99**, 83–258.
285. Maze, J. (1968). Past hybridization between *Quercus macrocarpa* and *Q. gambelii*. *Brittonia* **20**, 321–333.
286. Melville, R. (1966). Continental drift: Mesozoic continents and the migration of angiosperms. *Nature (London)* **211**, 116–120.
287. Meyer, E. R. (1973). Late-Quaternary paleoecology of the Cuatro Ciénegas Basin, Coahuila. *Ecology* **54**, 982–995.
288. Miranda, F., and Sharp, A. J. (1950). Characteristics of the vegetation in certain temperate regions of eastern Mexico. *Ecology* **31**, 313–333.
289. Moir, W. H. (1969). Steppe communities in the foothills of the Colorado Front Range and their relative productivities. *Am. Midl. Nat.* **81**, 331–340.
290. Moir, W. H. (1969). The lodgepole pine zone in Colorado. *Am. Midl. Nat.* **81**, 87–98.
291. Monk, C. D. (1965). Southern mixed hardwood forest of north central Florida. *Ecol. Monogr.* **35**, 335–354.
292. Monk, C. D. (1966). An ecological study of hardwood swamps in north-central Florida. *Ecology* **47**, 649–654.
293. Monk, C. D. (1968). Successional and environmental relationships of the forest vegetation of north central Florida. *Am. Midl. Nat.* **79**, 441–457.
294. Monk, C. D., and Brown, T. W. (1965). Ecological consideration of cypress heads in north central Florida. *Am. Midl. Nat.* **74**, 126–140.
295. Mooney, H. A., *et al.* (1970). Vegetation comparisons between the Mediterranean climatic areas of California and Chile. *Flora* **159**, 480–496.
296. Moreau, R. E. (1955). Ecological changes in the palaearctic region since the Pleistocene. *Proc. Zool. Soc. London* **125**, 253–295.
297. Morton, J. K. (1972). Phytogeography of the west African mountains. *In* "Taxonomy, Phytogeography and Evolution" (D. H. Valentine, ed.), pp. 221–236. Academic Press, New York.
298. Moss, E. H. (1955). The vegetation of Alberta. *Bot. Rev.* **21**, 493–567.
299. Moyle, J. B. (1946). Relict boreal plants of southeastern Minnesota. *Rhodora* **48**, 163.
300. Muller, C. H. (1939). Relations of the vegetation and climatic types in Nuevo Leon, Mexico. *Am. Midl. Nat.* **21**, 687–729.
301. Muller, C. H. (1947). Vegetation and climate of Coahuila, Mexico. *Madrono* **9**, 33–57.
302. Muller, J. (1970). Palynological evidence on early differentiation of angiosperms. *Biol. Rev. Cambridge Philos. Soc.* **45**, 417–450.
303. Myers, C. W. (1969). The ecological geography of cloud forest in Panama. *Am. Mus. Novit.* No. 2396, 1–52.
304. Nevling, L. I., Jr. (1971). The ecology of an elfin forest in Puerto Rico. 16. The flowering cycle and an interpretation of its seasonality. *J. Arnold Arbor, Harvard Univ.* **52**, 586–613.
305. Nichol, A. A. (1952). The natural vegetation of Arizona. *Ariz. Agric. Exp. Stn., Tech. Bull.* **127**, 187–230.
306. Noy-Meir, I. (1973). Desert ecosystems: environment and producers. *Annu. Rev. Ecol. Syst.* **4**, 25–51.
307. Olmsted, C. E. (1945). Growth and development in range grasses. V. Photoperiodic responses of clonal divisions of three latitudinal strains of side-oats grama. *Bot. Gaz.* **106**, 382–401.
308. Oosting, H. J. (1954). Ecological processes and vegetation of the maritime strand in the southeastern United States. *Bot. Rev.* **20**, 226–262.

309. Oosting, H. J., and Billings, W. D. (1943). The red fir forest of the Sierra Nevada. *Ecol. Monogr.* **13**, 259–274.

310. Oosting, H. J., and Billings, W. D. (1951). A comparison of virgin spruce-fir forest in the northern and southern Appalachian system. *Ecology* **32**, 84–104.

311. Opdyke, N. D. (1961). The paleoclimatological significance of desert sandstone. *In* "Descriptive Paleoclimatology" (A. E. M. Nairn, ed.), pp. 45–60. Wiley (Interscience), New York.

312. Pady, S. M., and Kapica, L. (1955). Fungi in air over the Atlantic Ocean. *Mycologia* **47**, 34–50.

313. Palmer, W. C., and Havens, A. V. (1958). A graphical technique for determining evapotranspiration by the Thornthwaite technique. *Mon. Weather Rev.* **86**, 123–128.

314. Parsons, R. F., and Cameron, D. G. (1974). Maximum plant species diversity in terrestrial communities. *Biotropica* **6**, 202–203.

315. Pase, C. P., and Lindemuth, A. W., Jr. (1971). Effects of prescribed fire on vegetation and sediment in oak-mountain mahogany chaparral. *J. For.* **69**, 800–805.

316. Payson, E. B. (1922). A monograph of the genus *Lesquerella*. *Ann. Mo. Bot. Gard.* **8**, 103–236.

317. Penfound, W. T. (1952). Southern swamps and marshes. *Bot. Rev.* **18**, 413–446.

318. Penland, W. C. (1941). Aspects of the páramo of Ecuador. *Colo.-Wyo. Acad. Sci.* **3**, 35–36.

319. Pequegnat, W. E. (1951). The biota of the Santa Ana Mountains. *J. Entomol. Zool.* **42**(3/4), 1–84.

320. Phillips, W. S. (1963). Depth of roots in soil. *Ecology* **44**, 424.

321. Polunin, N. (1948). Botany of the Canadian eastern Arctic. Part III. Vegetation and ecology. *Natl. Mus. Can., Bull.* **104**, 304 pp.

322. Pond, F. W., and Cable, D. R. (1960). Effect of heat treatment on sprout production of some shrubs of the chaparral in central Arizona. *J. Range Manage.* **13**, 313–317.

323. Poore, M. E. D. (1963). Problems in the classification of tropical rain forest. *J. Trop. Geogr.* **17**, 12–19.

324. Poore, M. E. D. (1968). Studies in Malaysian rain forest. I. The forest on Triassic sediments in Jungka Forest Reserve. *J. Ecol.* **56**, 143–196.

325. Potter, L. D., and Greene, D. L. (1964). Ecology of a northeastern outlying stand of *Pinus flexilis*. *Ecology* **45**, 866–868.

326. Quarterman, E. (1950). Major plant communities of Tennessee cedar glades. *Ecology* **31**, 234–255.

327. Quarterman, E., and Keever, C. (1962). Southern mixed hardwood forest: climax in the southeastern coastal plain, U.S.A. *Ecol. Monogr.* **32**, 167–185.

328. Quarterman, E., *et al.* (1972). Analysis of virgin mixed mesophytic forests in Savage Gulf, Tennessee. *Bull. Torrey Bot. Club* **99**, 228–232.

329. Raup, H. M. (1940). Old field forests of southeastern New England. *J. Arnold Arbor, Harvard Univ.* **21**, 266–273.

330. Raven, P. H. (1963). Amphitropical relationships in the floras of North and South America. *Q. Rev. Biol.* **38**, 151–177.

331. Raven, P. H., and Axelrod, D. I. (1972). Plate tectonics and Australasian paleobiogeography. *Science* **176**, 1379–1386.

332. Raven, P. H., and Axelrod, D. I. (1975). History of the flora and fauna of Latin America. *Am. Sci.* **63**, 420–429.

333. Reed, E. C., *et al.* (1965). The Pleistocene in Nebraska and northern Kansas. *In* "The Quaternary of the United States" (H. E. Wright, Jr. and D. G. Frey, eds.), pp. 187–202. Princeton Univ. Press, Princeton, New Jersey.

334. Reed, R. M. (1971). Aspen forests of the Wind River Mountains, Wyoming. *Am. Midl. Nat.* **86**, 327–343.

335. Reid, E. M., and Chandler, M. E. J. (1933). "The London Clay Flora," 561 pp. *Br. Mus. Nat. His.*, London.
336. Retzer, J. L. (1956). Alpine soils of the Rocky Mountains. *J. Soil Sci.* **7**, 22–32.
337. Rice, R. M., and Foggin, G. T., III (1971). Effect of high intensity storms on soil slippage on mountainous watersheds in southern California. *Water Resour. Res.* **7**, 1485–1496.
338. Richards, P. W. (1956). Study of tropical vegetation with special reference to British Guiana and British West Africa. *Unasylva* **10**, 161–165.
339. Richards, P. W. (1964). "The Tropical Rain Forest," 423 pp. Cambridge Univ. Press, London and New York.
340. Richards, P. W. (1969). Speciation in the tropical rain forest and the concept of the niche. *J. Ecol.* **57**, 3P–4P.
341. Richmond, G. M. (1965). Glaciation of the Rocky Mountains. *In* "The Quaternary of the United States" (H. E. Wright, Jr. and D. G. Frey, eds.), pp. 217–230. Princeton Univ. Press, Princeton, New Jersey.
342. Rickard, W. H., and Beatley, J. C. (1965). Canopy-coverage of the desert shrub vegetation mosaic of the Nevada Test Site. *Ecology* **46**, 524–529.
343. Rigg, G. B. (1937). Some raised bogs of southeastern Alaska with notes on flat bogs and muskegs. *Am. J. Bot.* **24**, 194–198.
344. Ritchie, J. C., and Hare, F. K. (1971). Late-Quaternary vegetation and climate near the Arctic tree line of northwestern North America. *Quat. Res. (N.Y.)* **1**, 331–342.
345. Rizzini, C. T., and Ezechias, P. H. (1961). Underground organs of plants from some southern Brazilian savannas, with special reference to the xylopodium. *Phyton* **17**, 105–124.
346. Rosseau, J. (1948). The vegetation and life zones of George River, eastern Ungava, and the welfare of the natives. *Arctic* **1**, 93–96.
347. Rune, O. (1954). Notes on the flora of the Gaspé Peninsula. *Sven. Bot. Tidskr.* **48**, 117–136.
348. Russell, N. H. (1953). The beech gaps of the Great Smoky Mountains. *Ecology* **34**, 366–374.
349. Rzedowski, J. (1954). (Vegetation of the San Angel Pedregal (D.F., Mexico)). *An. Esc. Nac. Cienc. Biol., Mexico City* **8**, 59–129.
350. Sampson, A. W. (1944). Plant succession on burned chaparral lands in northern California. *Calif. Agric. Exp. Stn., Bull.* **685**, 144 pp.
351. Savile, D. B. O. (1972). Arctic adaptations in plants. *Can. Dep. Agric., Res. Branch, Monogr.* **6**, 81 pp.
352. Scheer, B. T. (1945). The development of marine fouling communities. *Biol. Bull.* **89**, 103–121.
353. Schuchert, C. (1955). "Atlas of Paleogeographic Maps of North America," 177 pp. Wiley, New York.
354. Schulman, E. (1954). Longevity under adversity of conifers. *Science* **19**, 396–399.
355. Schulz, J. P. (1960). Ecological studies on rain forest in northern Suriname. *Vehr. K. Ned. Akad. Wet., Afd. Natuurk.* **53**, 267 pp.
356. Sears, P. B., *et al.* (1955). Palynology in southern North America. Parts 2 and 3. *Geol. Soc. Am. Bull.* **66**, 471–530.
357. Sellers, W. D. (1965). "Physical Climatology," 272 pp. Univ. of Chicago Press, Chicago, Illinois.
358. Shanks, R. E. (1954). Climates of the Great Smoky Mountains. *Ecology* **35**, 354–361.
359. Shantz, H. L., and Zon, R. (1924). "Natural Vegetation," Atlas Am. Agric., Part 1, Sect. E, pp. 15–19. U.S. Dep. Agric., Washington, D.C.
360. Sharp, A. J., and Iwatsuki, Z. (1965). A preliminary statement concerning mosses common to Japan and Mexico. *Ann. Mo. Bot. Gard.* **52**, 452–456.
361. Shaw, C. H. (1916). The vegetation of the Selkirks. *Bot. Gaz.* **61**, 477–494.
362. Shreve, F. (1914). A montane rain forest. *Carnegie Inst. Washington Publ.* **199**, 110 pp.

363. Shreve, F. (1937). Lowland vegetation of Sinaloa. *Bull. Torrey Bot. Club* **64**, 605–613.

364. Shreve, F. (1937). The vegetation of the Cape Region of Baja California. *Madrono* **4**, 105–113.

365. Shreve, F. (1942). The desert vegetation of North America. *Bot. Rev.* **8**, 195–246.

366. Shreve, F. (1964). Vegetation and Flora of the Sonoran Desert, Vol. 1, Vegetation of the Sonoran Desert. *Carnegie Inst. Washington Publ.* **591**, 192 pp.

367. Siccama, T. G. (1974). Vegetation, soil and climate on the Green Mountains of Vermont. *Ecol. Mono.* **44**, 325–349.

368. Silverberg, R. (1967). "The World of the Rain Forest, 172 pp. Appleton (Meredith), New York.

369. Simpson, G. G. (1952). Probabilities of dispersal in geologic time. *Bull. Am. Mus. Nat. Hist.* **99**, 163–176.

370. Sivarajasingham, L. T., *et al.* (1962). Laterite. *Adv. Agron.* **14**, 1–60.

371. Skottsberg, C. (1938). Geographical isolation as a factor in species formation, and its relation to certain insular floras. *Proc. Linn. Soc. London* **150**, 286–293.

372. Sloan, R. E. (1969). Cretaceous and Paleocene terrestrial communities of western North America. *Proc. North Am. Paleontol. Conf.* **1**, 427–453.

373. Smiley, C. J. (1966). Cretaceous floras from Kuk River area, Alaska: Stratigraphic and climatic interpretations. *Geol. Soc. Am. Bull.* **77**, 1–14.

374. Smith, A. C. (1973). Angiosperm evolution and the relationship of the floras of Africa and America. *In* "Tropical Forest Ecosystems in Africa and South America" (B. J. Meggers, *et al.*, eds.), pp. 49–61. Random House (Smithsonian Inst. Press), New York.

375. Soerianegara, I. (1971). Characteristics and classification of mangrove soils of Java. *Rimba Indones.* **16**, 141–150.

376. Solbrig, O. T. (1972). New approaches to the study of disjunctions with special emphasis on the American amphitropical desert disjunctions. *In* "Taxonomy, Phytogeography and Evolution" (D. H. Valentine, ed.), pp. 85–100. Academic Press, New York.

377. Sorenson, T. (1941). Temperature relations and phenology of the northeast Greenland flowering plants. *Medd. Groenl.* **125**(9), 1–302.

378. Sparling, J. H. (1967). Assimilation rates of some woodland herbs in Ontario. *Bot. Gaz.* **128**, 160–168.

379. Specht, R. L. (1969). A comparison of the sclerophyllous vegetation characteristic of Mediterranean type climates in France, California, and southern Australia. *Aust. J. Bot.* **17**, 277–308.

380. Spetzman, L. A. (1959). Vegetation of the Arctic slope of Alaska. *U.S. Geol. Surv. Prof. Pap.* **302**, 19–58.

381. Stanford, W. W. (1969). The distribution of epiphytic orchids in relation to each other and to the geographic location and climate. *Biol. J. Linn. Soc. London*, **1**, 247–285.

382. Stearns, F. (1951). The composition of the sugar maple-hemlock-yellow birch association in northern Wisconsin. *Ecology* **32**, 245–265.

383. Stephens, F. R., *et al.* (1970). The muskegs of southeast Alaska and their diminished extent. *Northwest Sci.* **44**, 123–130.

384. Stephenson, T. A., and Stephenson, A. (1972). "Life between Tidemarks on Rocky Shores," 425 pp. Freeman, San Francisco, California.

385. Stern, W. L., and Eyde, R. H. (1963). Fossil forests of Ocu, Panama. *Science* **140**, 1214.

386. Steyermark, J. A. (1950). Flora of Guatemala. *Ecology* **31**, 368–372.

387. Takhtajan, A. (1969). "Flowering Plants: Origin and Dispersal," 310 pp. Oliver & Boyd, Edinburgh.

388. Taylor, B. W. (1954). An example of long distance dispersal. *Ecology* **35**, 569–572.

389. Tedrow, J. C. F. (1966). Polar desert soils. *Soil Sci. Soc. Am., Proc.* **30**, 381–387.

390. Tedrow, J. C. F., and Harries, H. (1960). Tundra soil in relation to vegetation, permafrost and glaciation. *Oikos* **11**, 237–249.

391. Thorne, R. F. (1973). Floristic relationships between tropical Africa and tropical America.

In "Tropical Forest Ecosystems in Africa and South America" (B. J. Meggers, *et al.*, eds.), pp. 27–48. Random House (Smithsonian Inst. Press), New York.

392. Thornthwaite, C. W. (1943). Problems in the classification of climates. *Geogr. Rev.* **33**, 233–255.

393. Transeau, E. N. (1935). The prairie peninsula. *Ecology* **16**, 423–437.

394. Turesson, G. (1922). The species and variety as ecological units. *Hereditas* **3**, 100–113.

395. Turner, R. M. (1963). Growth in four species of Sonoran Desert trees. *Ecology* **44**, 760–765.

396. UNESCO (1958). *Study of Trop. Veg.: Proc. Kandy Symp.*, Paris 226 pp.

397. Ungar, I. A. (1965). An ecological study of the vegetation of the Big Salt Marsh, Stafford County, Kansas. *Univ. Kans. Sci. Bull.* **46**, 1–99.

398. Valentine, J. W., and Moores, E. M. (1970). Plate-tectonic regulation of faunal diversity and sea level: a model. *Nature (London)* **228**, 657–659.

399. Van Houten, F. B. (1945). Review of latest Paleocene and early Eocene mammalian faunas. *J. Paleontol.* **19**, 421–461.

400. Van Overbeek, J., and Crist, R. E. (1947). The role of a tropical green alga in beach sand formation. *Am. J. Bot.* **34**, 299–300.

401. Van Ryswyk, A. L., *et al.* (1966). The climate, native vegetation, and soils of some grasslands at different elevations in British Columbia. *Can. J. Plant Sci.* **46**, 35–50.

402. van Steenis, C. G. G. J. (1957). Outline of vegetation types in Indonesia and some adjacent regions. *Pac. Sci. Congr. Proc., 8th* **4**, 61–97.

403. van Steenis, C. G. G. J. (1959). Tropical lowland vegetation: the characteristics of its types and their relation to climate. *Pac. Sci. Congr. Proc., 9th* **20**, 25–37.

404. van Steenis, C. G. G. J. (1962). The mountain flora of the Malaysian tropics. *Endeavour* **21**, 183–193.

405. Vavilov, N. I. (1941). "The Origin, Variation, Immunity and Breeding of Cultivated Plants," 364 pp. Chron. Bot., Waltham, Massachusetts.

406. Vogelmann, H. W. (1973). Fog precipitation in the cloud forests of eastern Mexico. *BioScience* **23**, 96–100.

407. Vogl, R. J., and McHargue, L. T. (1966). Vegetation of California fan palm oases on the San Andreas fault. *Ecology* **47**, 532–540.

408. Vogl, R. J., and Schorr, P. K. (1972). Fire and manzanita chaparral in the San Jacinto Mountains, California. *Ecology* **53**, 1179–1188.

409. Vuilleumier, B. S. (1971). Pleistocene changes in the fauna and flora of South America. *Science* **173**, 771–780.

410. Vuilleumier, F. (1970). Insular biogeography in continental regions. I. The northern Andes of South America. *Am. Nat.* **104**, 373–388.

411. Waibel, L. (1948). Vegetation and land use in the planalto of central Brazil. *Geogr. Rev.* **38**, 529–544.

412. Walker, E. H., and Pendleton, R. L. (1957). A survey of the vegetation of southeastern Asia, the Indo-Chinese Province of the Pacific Basin. *Pac. Sci. Congr. Proc., 8th* **4**, 99–113.

413. Wallace, A. R. (1902). "Island Life," 2nd Ed., 563 pp. Macmillan, New York.

414. Walter, H. (1973). "Vegetation of the Earth in Relation to Climate and the Eco-Physiological Conditions," 240 pp. Springer-Verlag, Berlin and New York.

415. Watts, W. A. (1971). Postglacial and interglacial vegetation history of southern Georgia and central Florida. *Ecology* **52**, 676–690.

416. Weaver, H. (1968). Fire and its relationship to ponderosa pine. *Proc. Tall Timbers Conf. Ecol. Anim. Control Habitat Manage.* **7**, 127–150.

417. Weaver, J. E. (1924). Plant production as a measure of environment. A study in crop ecology. *J. Ecol.* **12**, 205–237.

418. Weaver, J. E., and Albertson, F. W. (1936). Effects of the great drought on the prairies of Iowa, Nebraska and Kansas. *Ecology* **17**, 567–639.

419. Weaver, J. E., and Albertson, F. W. (1956). "Grasslands of the Great Plains," 395 pp. Johnsen Publ. Co., Lincoln, Nebraska.
420. Weaver, J. E., and Clements, F. E. (1938). "Plant Ecology," 2nd Ed., 601 pp. McGraw-Hill, New York.
421. Webster, J. D. (1950). Altitudinal zonation of birds in southeastern Alaska. *Murrelet* **31**, 23–26.
422. Wellman, F. L. (1968). More diseases on crops in the tropics than in the temperate zone. *Ceiba* **14**, 17–28.
423. West, N. E., and Ibrahim, K. I. (1968). Soil-vegetation relationships in the shadscale zone of southwestern Utah. *Ecology* **49**, 445–456.
424. West, R. C. (1957). The Pacific lowlands of Colombia. *La. State Univ. Stud., Sci. Ser.* **8**, 278 pp.
425. Whitfield, C. J., and Anderson, H. L. (1938). Secondary succession of the desert plains grassland. *Ecology* **19**, 171–180.
426. Whitford, H. N. (1906). The vegetation of the Lamao Forest Reserve. *Philipp. J. Sci.* **1**, 373–431, 637–682.
427. Whitman, W. C., *et al.* (1943). Natural revegetation of abandoned fields in western North Dakota. *N.D. Agric. Exp. Stn., Bull* **321**, 18 pp.
428. Wilken, G. C. (1967). History and fire record of a timberland brushfield in the Sierra Nevada of California. *Ecology* **48**, 302–304.
429. Willett, H. C. (1950). Temperature trends of the past century. *Proc. R. Meteorol. Soc. Centen.* pp. 195–206.
430. Williams, L. (1967). Forests of southeast Asia, Puerto Rico, and Texas. *U.S. Dep. Agric., Agric. Res. Serv., Crops Res. Div. CR* **12–67**, 410 pp.
431. Willis, J. C. (1922). "Age and Area: A Study in Geographical Distribution and Origin," 259 pp. Univ. Press, Cambridge, England.
432. Wilson, L. R., and Webster, R. M. (1946). Plant microfossils from a Fort Union coal of Montana. *Am. J. Bot.* **33**, 271–278.
433. Wimpenny, R. S. (1966). "The Plankton of the Sea," 426 pp. Elsevier, New York.
434. Wolfe, J. A. (1968). Paleogene biostratigraphy of nonmarine rocks in King County, Washington. *U.S. Geol. Surv., Prof. Pap.* **571**, 33 pp.
435. Wolfe, J. A. (1972). An interpretation of Alaskan Tertiary floras. *In* "Floristics and Paleofloristics of Asia and Eastern North America" (A. Graham, ed.), pp. 201–233. Elsevier, New York.
436. Wolfe, J. A., and Hopkins, D. M. (1967). Climatic changes recorded by Tertiary land floras in northwestern North America. *In* "Tertiary Correlation and Climatic Changes in the Pacific" (K. Hatai, ed.), pp. 67–76. Saski, Sendai, Japan.
437. Wolfe, J. A. and Leopold, E. B. (1967). Neogene and early Quaternary vegetation of northwestern North America and northeastern Asia. *In* "The Bering Land Bridge" (D. M. Hopkins, ed.), pp. 193–206. Stanford Univ. Press, Stanford, California.
438. Wright, H. E., Jr. (1971). Late Quaternary vegetational history of North America. *In* "Late Cenozoic Glacial Ages" (K. K. Turekian, ed.), pp. 425–464. Yale Univ. Press, New Haven, Connecticut.
439. Wyatt-Smith, J. (1957). Peat swamp forest in Malaya, *Malays. For.* **22**, 5–32.
440. Wynne-Edwards, V. C. (1937). Isolated arctic-alpine floras in eastern North America: a discussion of their glacial and recent history. *Trans. R. Soc. Can. Sect. 5* **31**, 1–26.
441. Yendo, K. (1914). On the cultivation of seaweeds, with special accounts of their ecology. *Econ. Proc. R. Dublin Soc.* **2**, 105–122.
442. Zach, L. W. (1950). A northern climax, forest or muskeg? *Ecology* **31**, 304–307.

GLOSSARY

Aestivate   To pass the summer in a dormant condition.

Alien   A species occurring in an area to which it is not native, i.e., one introduced very recently, and usually by man.

Allele   One member of a pair of genes that occur in the same relative position in homologous chromosomes, and influences the manner of expression of a particular character.

Allopatric   Referring to taxa which occupy mutually exclusive ranges.

Alpine   Those portions of mountains that rise above the cold limits of trees.

Alpine Meadow Soils   Soil profiles that are usually shallow and stony, with very thin litter and duff, a dark grayish–brown to black A horizon which is 15–30 cm thick, strongly acid and granular to crumb structured, with a B horizon lacking, and a C horizon gleyed owing to poor drainage.

Alpine Turf Soils   Soil profiles with very thin litter and duff, a black A horizon that is strongly acid, stony and 10–30 cm thick, a dark yellowish–brown B horizon that is 15–30 cm thick, stony and medium acid. The profile is well drained, but sometimes subject to frost churning.

Antarctic   Pertaining to unforested areas that lie mostly south of the Antarctic Circle at 23°30′ from the South Pole.

Aphotic   Referring to depths in lakes and seas at and below which light becomes too feeble to affect living organisms.

Arctic   Pertaining to unforested areas that lies mostly north of the Arctic Circle at 23°30′ from the North Pole.

Arctic–alpine   Pertaining to arctic and alpine regions jointly.

Arctic Brown Soils   Soil profiles usually no more than 15 cm thick, with a yellowish–brown A horizon that is single grain or crumb structured and acid, lying directly on a C horizon that is circumneutral.

Association, Plant   A kind of plant community represented by stands occurring in places where environments are so closely similar that there is a high degree of floristic uniformity in all layers.

Austral   Southerly; the antonym of boreal.

Autecology   That branch of biology which deals with the interrelationships between the individual organism or species and its environment.

Bajada   Syn.: pediment. The elevated but gentle outwash slope at the base of a mountain in an arid landscape, typically with coarse stony soils.

Barrial Mud flat. Synonymous with playa, frequently, but more precise in its implication.

Barrier Any tract of inhospitable space broader than the probable range of dissemination, this preventing a species from extending its range.

Beach Syn.: playa. That strip of land along the margin of a body of water where wave action is strong enough to inhibit all or nearly all vegetation.

Benthon Organisms attached to or living in close proximity to the bottoms of seas or lakes.

Bicentric Pertaining to a range that is divided between just two areas that are separated by a barrier.

Biotype All individuals of a species or subspecies that seem genetically identical. An ecotype usually consists of a multitude of biotypes, all of which have closely similar habitat requirements, but differ in other respects.

Bipolar Pertaining to a bicentric range separated by the equatorial belt.

Bog A class of oligotrophic mires usually replete with *Sphagnum*.

Boreal Northern; the antonym of austral.

Brown Forest Soils Soil profiles with negligible litter and duff, a very dark brown A horizon that is friable, a lighter colored B horizon that is slightly blocky, with the entire profile circumneutral as a result of the high lime content of the parent materials.

Brown Soils Soil profiles with a dark brown to dark grayish brown A horizon that is 20–25 cm thick, platy above and granular below, slightly alkaline but not calcareous, a B horizon that is yellowish brown, 10–50 cm thick, weakly subangular blocky to prismatic, medium alkaline, and a lime layer in the lower B or upper C horizon.

Brown Podzolic Soils Soil profiles with a moderate covering of litter and duff, a dark grayish– or yellowish–brown A horizon that is 2–5 cm thick, usually with a thin and intermittent bleached layer in the lower part, a brown or yellow–brown B horizon that is 10–50 cm thick, single-grained, granular or slightly blocky but with no appreciable increase in clay, and medium to strongly acid.

Bulbil A specialized bud produced above ground, which serves as a reproductive organ when it becomes detached.

Bunchgrass A perennial herbaceous grass lacking rhizomes. A caespitose or tussock grass.

Calcification A soil-forming process confined to regions where precipitation is sufficient to leach the easily weathered products of primary minerals, principally $CaCO_3$, only a short distance down in the soil profile before they are precipitated. The resultant lime layer occurs in either the lower B or upper C horizon. The dominant clay formed in weathering is montmorillonite, which has a very high cation exchange capacity.

Caliche A North American colloquial term for the more or less cemented layer of $CaCO_3$, with or without an admixture of $MgCO_3$, which is formed during the calcification of soil profiles.

Carr Mire supporting scrub, but too well supplied with basic cations to support *Sphagnum*.

Cauliflory A character of woody plants, chiefly small tropical trees, in which the flowers are produced on old wood (branches or trunk) rather than on twigs of the current season.

Chamaephyte Dwarf shrubs with no live buds more than 25 cm above the ground surface during the most unfavorable season.

Chaparral A rather closed cover of shrubs, mostly hard-wooded and evergreen, that is restricted to areas with hot summers with deficient precipitation, and winters with at least occasional frost.

Chernozem Soils Soil profiles with a very dark brown or black A horizon that is 15–50 cm thick, moderately prismatic to blocky, and with a lime layer in the lower B or upper C horizon.

Chestnut Soils Soil profiles with a dark grayish–brown to very dark brown A horizon that is 10–45 cm thick, granular or platy and slightly alkaline, and a brown B horizon that is 15–75 cm thick, weakly prismatic to subangular blocky, and with a lime layer in the lower B or upper C horizon.

Circumboreal Syn.: Holarctic. Occurring at once in the northern parts of North America, Asia and Europe.

Climatic climax The apparently stable vegetation that terminates succession on zonal soils.

Climax That state of a biotic community that is attained when population structures of all its species fluctuate rather than exhibit unidirectional change. Such a community will remain in a self-perpetuating state so long as present climatic, edaphic and biotic conditions continue. Antonym of seral.

Clone A population derived from a single parental plant by vegetative reproduction.

Community Any group of organisms interacting among themselves.

Compensation, Habitat of A restricted area of microclimate or soil which allows an organism or a community to live in an area generally inimical to its survival.

Competition The influence of one organism on another which results from both drawing on one or more resources that are in short supply.

Continental climate Any climate in which the difference between summer and winter temperatures is greater than average for that latitude in consequence of distance from a sea or ocean. Antonym of oceanic climate.

Convergent evolution Evolutionary development that results in genetically unrelated organisms becoming similar in structure and/or function.

Cosmopolitan A taxon native at once to all continents and widely distributed in each.

Coverage The area of ground surface included in a vertical projection of individual plant canopies.

Cryptogam Any plant reproducing sexually without forming seeds. Antonym of phanerogam.

Day-neutral Pertaining to plants which function normally under daily light periods ranging from much shorter to much longer than 12 hours.

Desert A category of ecosystems with high heat balance and insufficient rainfall to support a significant cover of perennial grass on zonal soils.

Desert soils Soil profiles with a pale gray or brownish–gray A horizon that is less than 15 cm thick, platy or vescicular, and a B horizon that is as dark or darker than the A, 15–35 cm thick, prismatic or blocky, usually with more clay, neutral to strongly alkaline, and with a lime layer, sometimes cemented, in the lower B or upper C horizon.

Diploid An organism containing only two sets of chromosomes in its cells, paternal and maternal.

Disclimax A distinctive type of climax community which retains its character only under continuous or intermittent disturbance such as heavy grazing or periodic burning. Exemplified by zootic and fire climaxes.

Disjunct Pertaining to a discontinuous range having two or more potentially interbreeding populations separated by a distance precluding genetic exchange by pollination or dissemination.

Dispersal Syn.: Dissemination.

Dissemination Syn.: Dispersal. The scattering of detached structures capable of reproducing a plant.

Disseminule A detached structure capable of reproducing a plant.

Dominance The collective size or bulk of the individuals of a group of organisms as it determines their relative influence on other components of the ecosystem.

Dominant A taxon or group of taxa which has high dominance.

Duff Decomposition products of litter lying on mineral soil, in which the identity of the original tissue can no longer be discerned. A product of litter decay.

Dystrophic Waters of ponds or streams that are brownish as a result of suspended organic colloids.

Ecocline A series of populations distributed along an environmental gradient, with each population exhibiting small difference in its genetic makeup which adapt it to its own segment of the environmental gradient.

Ecologic amplitude The breadth of environmental requirements and tolerances of a taxon.

Ecology   The science which deals with the reciprocal relations between organisms and their environment.

Ecosystem   An aggregation of organisms of any size, considered together with their physical environment.

Ecotone   Any zone of intergradation or interfingering, narrow or broad, between contiguous types of vegetation.

Ecotype   A race within a species which is genetically adapted to a habitat type that is different from the habitat types of other races of that species.

Edaphic   Pertaining to soil.

Edaphic climax   Any distinctive type of stable community that develops on soils different from those supporting a climatic climax.

Element, Floristic   A group of plants sharing a distribution pattern, a migratory history, or some other common feature important in plant geography.

Emigration   The migration of a taxon from a place in which it has been residing.

Endemic   An organism, or pertaining to an organism, which is restricted to a stated portion of the earth's surface.

Environment   The surroundings of an organism or a community.

Epiphyll   A plant growing perched upon another plant without damaging it, except perhaps in a mechanical manner.

Eutrophic   Pertaining to water which is fairly well supplied with the mineral nutrients required by green plants.

Evapotranspiration   The loss of water from transpiring plants plus the soil supporting them, expressed in terms of equivalent rainfall.

Evolution   Genetic change within a population or species through time.

Fell-field   A type of tundra ecosystem characterized by rather flat relief, very stony soil, and low, widely spaced vascular plants.

Fen   A class of eutrophic mires lacking *Sphagnum*, with graminoids dominant.

Fire climax   Any type of apparently stable vegetation whose distinctiveness depends on being burned at rather regular intervals.

Flora   A list of all plant species living in a particular area at a particular time. Although the flora involved is a basic character of any specific community, ordinarily the term is used for areas so large they include several to many types of communities.

Forb   An herb other than a grass, sedge or other plant with similar foliage.

Forest   An area of closely spaced trees having considerable extent, i.e., larger than what might be called a grove.

Gallery forest   A strip of forest closely confined to the margin of a stream that courses through a landscape otherwise unforested.

Gene   A portion of a chromosome that has the potentiality of influencing some character of an organism.

Geophyte   A perennial herb with its perennating bud(s) located well below the soil surface.

Glei (adj. Gleyed)   The grayish or bluish fine-textured material in a soil profile whose color is a consequence of poor aeration that reduced iron compounds. A gleyed horizon may contain pockets or streaks of red–brown soil where locally better drainage has allowed the oxidation of iron compounds in coarser material.

Graminoid   An herbaceous grass or plant of similar growth form.

Grassland   Technically only types of vegetation dominated by herbaceous graminoids, but sometimes loosely applied to any herb-dominated vegetation.

Gray–Brown Podzolic Soils   Soil profiles with very little litter and duff, with an A horizon 7–20 cm thick that is grayish brown to very dark gray above and light yellowish brown or pale yellow below, and medium acid, and a B horizon 25–75 cm thick that is darker than the A and has more clay, is subangular blocky and slightly to medium acid.

Gray Desert Soils   Soil profiles developing in climates warmer than where Sierozems occur, but not as warm as Red Desert Soil areas. Similar to Sierozems except for being calcareous even to the surface.

Gray Wooded Soils   Soil profiles with litter and duff 5–15 cm thick, a platy or vescicular A horizon 5–10 cm thick, with the upper part darkened by humus, the lower part light gray, slightly to medium acid, and a B horizon 15–90 cm thick that is blocky and medium acid to neutral.

Grove   A stand of forest or woodland of small extent, surrounded by lower vegetation or bare soil.

Growth form   A type of plant distinguished by size, morphology and duration of the vegetative body, irrespective of taxonomic or ecologic relations.

Guyot   A flat-topped seamount representing a former volcanic island in the ocean that was eroded to sea level before subsidence of the ocean floor carried it well below present sea level.

Habitat   The particular kind of environment in which a species or community is living.

Habitat type   A collective term for all parts of the land surface supporting or capable of supporting the same kind of climax plant association.

Halophyte   A plant usually found in soil or water too salty to be tolerated by the majority of plant species.

Hammock   A colloquial term used mainly in Florida for islandlike stands of dicot forest in a landscape covered mostly by pine forest.

Heliophyte   A plant that grows usually fully exposed to the sun. An *obligative* heliophyte *requires* full sun. A *facultative* heliophyte *can tolerate* full sun, but usually grows in shade.

Hemicryptophyte   An herb with its perennating bud(s) at the soil surface, so that during the most unfavorable season it (they) can get protection from only a cover of litter or snow.

Herb   Any plant that dies back to the ground surface each year, or, in the moist tropics, any plant with a soft, nonwoody stem.

Holarctic   Syn.: Circumboreal. Occurring at once in the northern parts of North America, Asia and Europe.

Homozygous   The condition in which the genes controlling a character have identical influence on that character. A completely homozygous individual could produce only one kind of gamete.

Horizon, soil   A horizontal component of a soil profile in which weathering and/or plant influences have had a distinctive influence, as evident in the field.

Humus   The finely divided, amorphous organic matter that is diffused through mineral material in a soil profile.

Hybrid   An offspring whose parents are genetically unlike, or a community composed of elements drawn from two other distinctive communities.

Hydrophyte   A plant usually found growing in water, or in soil containing water well in excess of field capacity most of the time.

Hydrosere   A time (not space!) sequence of seral communities, or a continuum of variation, which starts with hydrophytes and eventually leads to a particular kind of climax.

Indigene   Any native member of a biota, i.e., any species which was not introduced into the biota by man.

Immigration   The entrance of a taxon into an area where it is not native.

Insular endemic   A taxon that has originated in a physically restricted area and has been unable to extend its range beyond that area.

Invasion   The entrance of an organism into an area where it was not formerly represented.

Isohyet   A line on a map connecting points of equal rainfall.

Isotherm   A line on a map connecting points of equal temperature.

Jungle   Seral vegetation, especially in the ever-wet tropics, that is dense and difficult for man to traverse.

Krummholz   The belt of discontinuous scrub or groveland at alpine timberlines, composed of species which have the genetic potential of the tree life form, but in this ecotonal belt are both strongly dwarfed and misshapen.

Lake   A body of water with negligible current which lacks evident vegetation in a peripheral belt centered on the shoreline. The barrenness is usually attributable to strong wave action, but sometimes primarily a result of bare rock.

Laterite   A product of laterization in which the profile develops a soft, cheeselike texture and hardens irreversibly upon drying. Old laterites consist of about 85% aluminum and/or iron oxides.

Laterization   A soil-forming process largely confined to the tropics, in which primary minerals are converted to kaolin, a clay with low cation adsorption capacity, and silica is converted to silicic acid then leached away leaving oxides and hydroxides of aluminum and iron to accumulate in place as a very deep layer.

Layer, vegetation   A structural component of a community that may be recognized as consisting of plants of approximately uniform and distinctive stature.

Liana   A plant rooted in the soil but incapable of supporting itself erect, and gaining height by climbing other plants.

Life form   A type of growth form of plants that appears to have special adaptive significance. Life forms are: therophytes, geophytes, hemicryptophytes, chamaephytes and phanerophytes.

Lignotuber   An irregular enlargement of the summit of a tap root plus outgrowths therefrom, which develops just below the soil surface on a woody plant that has sprouted repeatedly after having its shoots killed by fire.

Lithosol   A very stony and usually shallow soil, often without evident development of horizons.

Litter   Plant parts dropped on the soil surface so recently that the organ from which they originated can be discerned rather readily.

Littoral   That portion of a sea shore subject to alternate submergence and emergence by normal tides.

Macroclimate   Climatic conditions (temperature, precipitation, relative humidity and sunshine) as recorded approximately 1.5 m above level ground, so that the data are relatively uninfluenced by topography, the character of the soil, or by vegetation taller than a mowed lawn. Most data gathered by official weather stations are of this character to facilitate geographic comparisons.

Marsh   An ecosystem dominated by herbaceous plants, and with the soil saturated for long periods if not permanently, but without surface accumulations of peat.

Meadow–steppe   Steppe occurring in climates almost moist enough for forest, and containing an abundance of perennial forbs.

Microclimate   Any set of climatic conditions differing from the macroclimate, owing to closeness to the ground, to vegetation influences, to aspect, to cold air drainage, etc.

Microsere   A time sequence of communities, of small areal extent, that may be observed even in climax stands. Microseres involve such processes as the replacement of a large individual plant after it dies, the sequence of decomposers that follow each other in a unit of litter, the development of vegetation on an abandoned ant nest.

Middle America   Central America plus frost-free parts of Mexico.

Migration   In plant geography, the extension of a self-perpetuating population of a species into an area where it was not previously represented. Migration is also used in other senses in biology. It may refer to the vertical movements of plankton organisms that are repeated daily, or to either regular or erratic movements of animals from place to place with the passing of the seasons.

Mire   An ecosystem in which the rooting medium consists of wet peat.

Monsoon   Pertaining to a climatic pattern or vegetation existing in such a climate, where a cool dry season alternates with a hot wet season. In the original and narrow sense, applicable to only the Indo–Malayan area.

Montane   Pertaining to mountain slopes below the alpine belt.

Mycorhiza   A structure consisting of an intimate union of a fungus mycelium with the (usually) subterranean organ of a higher plant. The mycelia extend out into the substrate where they extract minerals that are passed back to the higher plant, whereas the latter provides energy-containing compounds for the fungus.

Mycotrophic   Pertaining to plants which depend on mycorhizae for satisfactory nutrition.

Mutation   An abrupt genetic change in a reproductive cell which, if it endures, results in a lineage that is permanently distinctive.

Naturalized   Said of an alien that continues to perpetuate itself after being introduced into a new area.

Neoendemic   An endemic with its restricted range attributable to only its recency of origin.

Noncalcic Brown Soils   Soil profiles with a brown to reddish brown A horizon 15–55 mm thick, that is massive and hard when dry, and a B horizon that is redder and has more clay, is 10–75 cm thick, is massive or has prismatic structure, often with hard silica-cemented layers included, and often with a limited lime accumulation in the lower part.

Nunatak   A hill or mountain high enough to have escaped being completely covered when glacial ice inundated the surrounding land.

Oceanic climate   Any climate in extratropical latitudes in which the difference between summer and winter temperatures is less than average for that latitude in consequence of proximity to a sea or ocean. Such climates have relatively long spring and summer seasons, and relatively short summers and winters.

Oligotrophic   Pertaining to water that is poorly supplied with the nutrients needed by plants.

Ortstein   A hard, iron-cemented part of the B horizon that may form in a strongly developed podzol profile.

Paleoendemic   An endemic that was once more widespread and has narrowly escaped annihilation by some change in the earth's surface or its climate.

Pantropical   A taxon represented at once in the tropical latitudes of South America, Africa, and Asia.

Phanerophyte   A shrub or tree. Divisible into Nanophanerophytes 0.25–2 m tall, Microphanerophytes 2–8 m tall, Mesophanerophytes 8–30 m tall, and Megaphanerophytes over 30 m tall.

Photic zone   The surface layer of water in a lake or sea, extending down as far as light intensity is sufficient to affect living organisms directly.

Photoperiod   That portion of each 24-hour day when sunlight is bright enough to affect the functions of living organisms.

Photoperiodism   The response of an organism to varying photoperiods.

Phreatophyte   A plant in an arid region that has roots extending into the capillary fringe above a water table.

Physiognomy   The superficial appearance of a mass of vegetation.

Phytomass   The fresh weight or dry weight of all living plants per unit land area or volume of water.

Pioneer   An organism getting established on a relatively or absolutely bare area where there is as yet little or no competition.

Plankton   The passively floating or weakly swimming minute organisms that live suspended in water.

Playa   A beach. A vague term referring to sandy ocean beaches as well as barrials of desertic areas.

Pneumatophore   A root specialized in structure or function in such a manner as to especially enhance gas exchange between submersed organs and the free atmosphere.

Podzolization   A soil-forming process characteristic of environments where precipitation is abundant, soils have good internal drainage, and the parent materials are low in lime. Under these conditions basic ions are mostly lost to streams, the profile becomes markedly acid, and compounds of Fe and Al as well as humus and clay colloids are leached from the A horizon (especially the A2) to accumulate in the B horizon. Silica becomes the chief residue left in the A2 horizon, imparting a light or even white color under strong podzolization, whereas the B horizon becomes brownish, reddish or yellowish owing to the abundance of iron oxides. The dominant clay formed in this process is illite, which has a medium cation exchange capacity.

Podzol Soils   Soil profiles with a markedly acid layer of litter and duff 10–40 cm thick, with a very thin upper part of the A horizon (the A1) that dark brown or black and a thicker A2 layer that is light colored, with a B horizon that is reddish– or yellowish–brown, 5–25 cm thick, blocky or prismatic, and commonly cemented as ortstein. The entire profile is strongly acid.

Polyphyletic   Referring to a genus or larger taxon whose members have come from different phyletic stocks, but through convergent evolution have been mistakenly considered a natural group, or to a community whose members have been drawn from different geographic source areas.

Polyploid   A condition in which each cell has three or more sets of chromosomes, rather than just one maternal and one paternal set as possessed by a diploid.

Polytopic   Referring to a situation in which the range of a species is disjunct, usually occurring in three or more areas separated by barriers.

Pond   A body of water with negligible current, that has vegetation extending without interruption from the surrounding elevated land into the water, i.e., lacking the barren, wave-beaten beach of a lake.

Prairie Soils   Soil profiles with a very dark brown to black A horizon 15–50 cm thick, that is granular or platy and slightly to medium acid, a yellowish–brown or grayish–brown B horizon 15–75 cm thick, that is prismatic or blocky and slightly acid to neutral, but lacking a layer of precipitated lime.

Preadaptation   Any genetic character of an organism which now has no survival value, but which would enable it to cope with an environment different from that in which it occurs.

Profile   The sequence of visually distinguishable horizons in a soil, or layers in a plant community, as seen in vertical section. In a soil profile there is usually an upper (A) horizon that has been leached of solutes or colloids that were deposited in a B horizon below, with the relatively unweathered material still deeper referred to as the C horizon.

Province   As defined for use in this book: A subdivision of a Region that is restricted to part of one continent, and distinguished by dominant species that have had a common past history. It has a more narrowly defined range of climate than the Region of which it is part. It is composed of Sections.

Rain forest   Forests of frost-free ever-wet climates.

Rain shadow   The dry area immediately to the leeward of a mountain mass which lies athwart air masses that move prevailingly in one direction across the mountain.

Range   That portion of the earth's surface enclosed by a line drawn about the outermost limits of the distribution of a taxon. An organism does not occupy all the area within its range owing to differences in soil, topography, etc.

Recessive   Pertaining to genes which result in the appearance of a character only when no dominant allele is present.

Red Desert Soils   Soil profiles in hot dry climates with a reddish or light reddish–brown A horizon that is single-grained, granular or platy, a B horizon 15–50 cm thick that is brighter red than the A, prismatic or blocky, weathering to become a clay pan, mildly to strongly alkaline, and with a lime layer in the lower B or upper C horizon.

Reddish Brown Soils   Variants of Brown Soils that develop in hotter climates, and have a distinct red or reddish–brown color.

Reddish–Brown Lateritic Soils   Soil profiles with thin litter and duff, a reddish– or yellowish–brown A horizon 15–50 cm thick, granular in the upper part, blocky below, often with shotlike magnesium concretions, slightly to medium acid, and a red to dark reddish–brown B horizon 35–150 cm thick, blocky, and strongly acid.

Reddish Chestnut Soils   Variants of Chestnut Soils that develop in hotter climates, have a reddish–brown A horizon and a reddish–brown to red B horizon.

Reddish Prairie Soils   Similar to Prairie Soils but developing in warm climates, and having a dark brown to reddish–brown A horizon, and a reddish–brown B horizon.

Red–Yellow Podzolic Soils   Soil profiles with little or no litter or duff, an A horizon 5–12 cm thick that is pale except near the surface, granular or blocky, medium to strongly acid, and a B horizon that is yellow or yellowish–red, 35–100 cm thick, blocky or prismatic and strongly acid. Since the clay in these soils is mostly kaolinite, they have low cation exchange capacity. Although approximately 50% of the solum is SiO_2, Fe and Al oxides remain well represented, as shown by the color.

Refugium   A small area in which organisms have survived when most of their former range became uninhabitable owing to climatic change or glaciation.

Region   As defined for use in this book: A collective term for all Provinces over the earth's surface which share a common physiognomy, and a grossly similar climatic pattern.

Relic   A small remnant of a population that was once widely distributed, or any taxon, irrespective of the size of its range, if it has outlived its close relatives.

Rendzina Soils   Soil profiles with a dark colored A horizon 15–50 cm thick, that is strongly calcareous at least in the lower part, granular or blocky, a B horizon which is blocky if present, and a lime layer in the lower B or upper C horizon. Calcareous rock fragments occur throughout the profile, and predominate below it.

Riparian   Pertaining to streamside environment.

Saline   Pertaining to soil or water containing sufficient soluble salts to be detrimental to the average plant.

Savanna   A physiognomic type of vegetation in which tall, widely spaced plants, especially trees, are scattered individually over land otherwise covered with low-growing plants, especially graminoids.

Sciophyte   A plant usually found growing in shaded environments.

Sclerophyllous   Pertaining to plants having thick, firm-textured leaves that are usually evergreen.

Scrub   Vegetation dominated by shrubs.

Seasonal   Pertaining to climates with alternating wet and dry seasons that are sufficiently different to affect the phenology of the vegetation as a whole.

Section   As defined for use in this book: A geographic subdivision of a Province with a measure of floristic distinctiveness. It consists of Zones.

Seral   Nonclimax, i.e., a species or a community demonstrably susceptible to replacement by another species or community, usually within a few centuries at most.

Sere   A sequence of communities that follow one another in the same habitat, and terminate in a particular kind of climax association.

Serotinous   Late in developing—a term used mainly for the cones of those *Coniferae* that remain closed for a year or more after the seeds within are fully mature.

Shrub–steppe   Steppe with a conspicuous shrub element, with the shrubs usually forming an open overstory above the grass layer.

Sierozem Soils   Soil profiles that are feebly differentiated, with a pale grayish A horizon less than 15 cm thick, platy and vesicular in the upper part, and a B horizon 15–35 cm thick, at least as dark as the A and browner or yellower, prismatic or blocky, mildly to strongly alkaline, usually calcareous, and with a lime layer in the lower B or upper C horizon.

Solum   A collective term for the A and B horizons of soil profile.

Stand   An uninterrupted unit of vegetation, homogeneous in composition and of the same age.

Steppe   Temperate zone vegetation dominated by grasses and occurring in climates where zonal soils are too dry to support trees.

Stratification   Pertaining to the grouping of plants in a community according to height classes.

Subalpine   The first distinctive type of vegetation, usually forest, below the alpine tundra, or a plant growing in such a location.

Succession   The partial or complete replacement of one community by another.

Swamp   An ecosystem dominated by woody plants and with soils saturated for long periods if not permanently, but without a surface accumulation of peat.

Symbiont   Either member of a pair of species that are directly and nutritionally interacting.

Sympatric   Referring to taxa which have overlapping ranges.

Synecology   The study of communities and their environmental relations.

Synusia   A subdivision of a plant community distinguished by its life form or microsite.

Taiga   Ecosystems adjacent to arctic tundra in which *Abies, Picea, Larix,* or paper-barked *Betula* are characteristic tree genera. Sometimes narrowly applied to just the arctic timberline ecotone; sometimes extended to all subarctic and even subalpine forests of the north temperate zone.

Taxon   A taxonomic group of any rank, such as subspecies, species, genus.

Temperate   Climates with regular winter seasons of freezing weather, alternating with summer seasons that are either hot, or only warm but of long duration.

Therophyte   An annual. A plant which completes its life cycle within a year.

Timberline   Any altitudinal or latitudinal limit of forest growth. Where high mountains rise from an arid basal plain there is both a lower and upper timberline.

Tropical   In an astronomic sense, the earth's surface between the Tropics of Cancer and Capricorn which lie 23½° from the poles of the earth. In a phytogeographic sense, climates with only rare and very light frosts at most.

Tundra   Ecosystems of areas beyond the cold limits of tree growth.

Tundra Soils   Soil profiles with 5–30 cm of litter and duff containing abundant roots and rhizomes, an A horizon about equally thick, that is dark brown and rests on a gleyed B horizon. The profile is low in clay content and base saturation, and is subject to frost churning which then destroys the horizonation.

Union   A subdivision of a plant association. It may be a single species of high abundance and distinctive ecology, or a rather well-defined list of species which are restricted to approximately the same narrow range of environmental variation in a vegetation mosaic. Commonly unions have physiognomic as well as taxonomic distinctiveness, i.e., they may consist of tall shrubs, or of herbs or of tree species, but this is not necessarily true. Therefore, union is a more flexible term than layer, emphasizing ecology as judged by similar patterns of distribution rather than height. The unions in a landscape occur in different combinations.

Vasculare   A vascular plant, i.e., fern or seed plant.

Vegetation   Plant life considered in mass.

Vicariad   A species or subspecies that has a close relative in another geographic area, with a barrier separating their ranges.

Weed   A plant growing where man does not want it to grow.

Woodland   Vegetation dominated by a rather closed stand of trees of short stature.

Western Brown Forest Soils   Soil profiles with litter and duff less than 5 cm thick, a dark A horizon 10–35 cm thick, granular and slightly to medium acid, and a B horizon that is 20–75 cm thick, subangular blocky, slightly to medium acid, and with a lime layer in the lower B or upper C horizon.

Xeromorphic   Having structural characteristics common among xerophytes, i.e., small thick leaves with sunken stomata or revolute margins, surfaces that are pubescent, waxy or highly reflective, and small vein islets. Xeromorphy is by no means restricted to xerophytes. It can be induced in mesophytes by low nutrient supplies, or it may be carried along in mesophytes or even hydrophytes that have evolved from xerophytes.

Zonal soils   Moderately deep to deep soil profiles developed from loamy parent materials, having moderate internal and surface drainage, and, except in extreme environments, with evident horizon differentiation. Soils in this category show the maximum correlation with climatic types.

Zonal vegetation   The vegetation of zonal soils.

Zone   As defined for use in this book: All the area in which zonal soils have the potentiality of supporting the same climatic climax plant association. It is composed of several to many habitat types, only one of which is the climatic climax.

Zootic climax   Any type of stable vegetation whose continued existence depends upon continued stress from heavy use by animals. The animal components of all ecosystems play important roles as subordinates, but only in a zootic climax is an animal so influential as to be clearly a dominant.

Index